History, Philosophy and Theory of the Life Sciences

Volume 12

More information about this series at http://www.springer.com/series/8916

Alvaro Moreno • Matteo Mossio

Biological Autonomy

A Philosophical and Theoretical Enquiry

Springer

Alvaro Moreno
IAS-Research Centre for Life,
Mind and Society
Departamento de Lógica y Filosofía
Universidad del País Vasco
Donostia – San Sebastian
Guipúzcoa, Spain

Matteo Mossio
Institut d'Histoire et de Philosophie des
Sciences et des Techniques (IHPST)
CNRS/Université Paris I/ENS
Paris, France

ISSN 2211-1948 ISSN 2211-1956 (electronic)
History, Philosophy and Theory of the Life Sciences
ISBN 978-94-017-9836-5 ISBN 978-94-017-9837-2 (eBook)
DOI 10.1007/978-94-017-9837-2

Library of Congress Control Number: 2015933996

Springer Dordrecht Heidelberg New York London

Printed on acid-free paper

Springer Science+Business Media B.V. Dordrecht is part of Springer Science+Business Media (www.springer.com)

Foreword

I am sitting with my grandchild in the park on a fading Australian summer afternoon. The sulphur-crested cockatoos screech as they squabble over the last of the sunlit eucalypt branches and she notices them as they fly over her, their powerful wings beating. "What is flying," she asks, "and why can't we do it?" For a moment I am tempted to respond to the latter question with "We don't have the genes for flight". But not only would this not help, indeed could not help, it also tears that question away from the initial one. While evolution as changes in gene frequencies can track the requisite gene changes involved, the actual features of organisms that make flight possible are left out. Genes can tell us about the appearance, spread and evolution of the *fact* of flight but they cannot in themselves tell us what flight actually *is*, namely the production of suitably spatially distributed and temporally coordinated thrust and lift. To understand that involves understanding, for example, how musculature must be recruited and organised to work wings that provide both lift and thrust, how skeletons must be both organised to effect tail-wing coordination, and be light enough to lift yet strong enough to brace the musculature in flight, to land on moving branches without fracturing legs, etc. and much more.

In short, it is to understand the internal organisation of birds. Without that we are blind to the internal consequences of genetic variation; and without that and ecological organisation, blind also to its external ecological consequences via the new sources of food and nests that become available, the spread of seeds via bird guts and the spread of plants that compete for bird feeding, and so on. Without such understanding the survival-of-the-fittest engine is left spinning its wheels, its simple idea of stochastic selection on populational variety left to sort rocks in a river and straws in the wind as well as gazelle on a savannah but without purchase on the nature and potential of evolving life.

There are three good reasons to read this book about how life is constituted. *First*, its organisational approach to organism is deeply informative, radically different from current orthodoxy and makes a crucial contribution at an important historical juncture in science. *Second*, it provides a detailed, powerful and ultimately elegant model of the mutual development of scientific and philosophical understanding. *Third*, the pellucid, penetrating and parsimonious character of the writing makes a text dense in precisely characterised ideas quite accessible, including to a non-expert audience.

The last of these features is uncommon, the second is decidedly unusual and the first quite unique. They are all discussed a little further below. If you have an interest in understanding how our world works, or a specific interest in the foundations of biological science and/or philosophy of biology, or in organised complex systems more widely (robotics, cybernetics, intelligent agents, etc.) then this book is for you.

The book expounds and explores the claim that a distinctive organisation is the hallmark of life and that organisation ultimately provides a framework for understanding the evolution of life forms, of agency and of intelligence/intentionality. A quick review of chapter content can be found towards the close of the Introduction. As you might expect, it starts with the basics, closure and self-maintenance, then a complex form of closure called autonomy, the foundational organisation of all life, and then explores the still more complex topics just noted. Moreno especially has pursued the organisational approach consistently over decades and, with Mossio as collegial co-writer, this book is the summative outcome. I have helped to make the odd contribution to this position myself, partly on its systems foundations (organisation), but largely concerned with the adaptive roots of cognition (see references, this book), and in my view this book is unique in offering the first high quality conceptually integrated, empirically grounded, in depth exposition of this approach. It shows just how far the organisational perspective can take us in understanding the nature and evolution of life (answer: very far) and its exposition bids fair to remain the standard for some time to come.

The Organisational Approach

Listing genes and gene-trait associations tells you little about how the creatures that carry the genes are put together. The common presumption is that those latter answers come after the genetic work is done and will be found by studying the biochemical detail. Then whatever organisation there is will drop out as a consequence. But there is another, reverse possibility, one that has been largely neglected, namely that there are irreducible structures of nested correlated interactions, that is, organisations, that are key to understanding why the biochemical details are as they are, genomes included, and that such organisational design is as fundamental to understanding as is the biochemistry. That is the approach taken here.

Organisation (think car engines) happens when many different parts (cylinders, cam shafts, fuel injection ...) interact in specific, coordinated ways (cylinder rod rotates on cam shaft, fuel injected into cylinder, ...) so as to collectively support some global functioning (convert chemical potential energy into torque). It is roughly measured by the numbers of nested layers of different correlations among different parts of a system. It is their functional contributions to the overall organisation of car engines that require the parts to have the shapes, sizes and material compositions they have. You can study these parts separately but unless you relate them to their organisation you will not understand their particular features. Organisation is not the same as order; pure crystals are highly ordered, so uniform they cannot show any organisation. Neither can gases, because they are too random to be organised. Organisation lies between the crystal and gas extremes but we don't have a good theory that tells us exactly where and why. Some may worry that talk of organisational constraints is too "airy fairy" and "metaphysical". But it is just the opposite, a matter of real dynamics found everywhere, from car engines to cellular "engines", for instance, the Krebs Cycle.

In this book the chief exemplar of an organised system is the living cell. The metabolism of a cell has to completely re-build the cell over time (that is its grand cycle). This is because, being material, a cell is a thermodynamic engine whose internal interactions degrade its innards which must then be replaced. But you don't get systematic self-replacement without being highly organised to do it: the particular materials and energy needed for each repair must be available at just the right location at just the right time, otherwise the cell will malfunction. In a cell more than 3,000 biochemical reactions are so organised that with each kind distinctively distributed throughout the cell their joint products re-make the cell, including themselves (and remove the thermodynamically unavoidable wastes), in the process also re-making the cell's capacity to extract from its environment the resources it needs. Thus at the heart of every cell is, and must be, a massive self-maintenance organisation cycle, operating under just the right constraints. This kind of organisation is called autonomy, with its core sense of self-governance applying all the way "up" from self-restriction by constraints to the more familiar socio-political notion.

Moreno and Mossio show that such organisation is central to cellular function, essentially defining all life. They also show that it is the necessary precursor to a well-defined evolutionary process, rather than the other way around. This is because the internal organisation of organisms secures the reproducibility of functionality which permits the inheritable traits, including those for mutant genomes, on which evolutionary selection operates. The interaction between evolutionary and developmental dynamics, in the context of epigenetic organisation, once mostly ignored but now richly studied, throws into stark relief the role of organism organisation in framing evolutionary process. All this is a relatively new perspective for evolutionary theorists, whose pure population statistics in themselves discourage awareness of organismal, communal and ecological organisation (cf. flight, above; albeit the theory has itself evolved significantly over the past 50 years). Moreno and Mossio lay out the issues with meticulous care.

Incidentally, it was the twin successes of the explorations of population genetics and molecular genetics that led to a century-long relative repression of biological organisation as an object of study, a repression that only really receded this century when molecular biology had exhausted simple gene sequencing and medicine simple gene-trait associations and both admitted the study of biosynthetic pathway organisation as the next major challenge. Thus this book arrives on the scene at this epochal moment, just in time to provide a penetrating framework for understanding what is actually involved in such research.

On that score, note that the science of spatio-temporal organisation of interactions so as to generate global self-maintenance is itself in its infancy; we know relatively little about it, but just enough about the incredibly complex ways reactants are spatially arranged in cells to suppose it is going to be a large, complex and very difficult domain to understand. But it must come if ever we are to develop a thorough cellular biology and much else up to truly life-like robotics beyond the one-dimensional computer-in-a-box toys we focus on at present. (See also Hooker ed. *Philosophy of Complex Systems*, North Holland 2011 for further discussion.)

Multicellular Organisation

The emergence of a biochemical organisation capable of regenerative closure, the cell, is the first decisive step in the evolution of life. A subsequent giant step is the organisation of groups of cells to form multicellular organisms. These must organise their multicellular processes so that cellular metabolism is supported throughout, hence the presence of a cardiovascular system to deliver oxygen and nutrients where needed and remove wastes, the presence of renal and lymph systems to manage toxins and so on. In short, multicellularity requires a set of "higher" organisational layers on top of cellular ones to obtain a functional organism. (Again, we do not as yet understand a lot about such organisational constraints, e.g. respiration, that reach from individual cells across organs and other intermediate organisations, to the whole organism.) But there is a pay-off for all this overhead.

The distinctive twin advantages of multicellularity lie in its increased capacities for more complex behaviours and for more interactively open organisation, each feeding the other, even while closure must still be satisfied for their component cells. Once cellular communication develops to allow cell specialisation compatibly with cellular organisational coherence (as above), the way is thrown open to great increases in both behavioural complexity and interactive openness. The case of expanded behavioural repertoires is obvious enough. No single cell can fly, for the good reason that, whether or not it can muster thrust, it cannot control its surface shape so as to provide lift. But a collection of cells suitably specialised and interconnected can provide the musculature, cardiovascular support, surface controllability and so on to fly, powerfully and elegantly.

The case of greater interaction openness is perhaps less obvious but of even greater significance. Multicellularity has made possible increases in interaction-led adaption of both inner metabolism and outer environment. In the case of inner metabolism, multicellular organisms are able to suspend or adapt aspects of metabolic activity, from speeding up some processes (e.g. removing wastes before conflict) to slowing down and modifying others (e.g. hibernation in bears), sometimes drastically (e.g. consuming internal organs for energy when fat stores are exhausted in stressful circumstances). Indeed, it is possible for existing organ systems to be entirely transformed in response to circumstances, as the metamorphosis of pupae into butterflies so beautifully illustrates. All of this requires over-arching organisational capacities. In the case of the outer environment, sensory cellular specialisation permits new ways of inward-bound interaction with the environment, leading to increased motor metabolic adaptiveness, from movement (e.g. sitting to running) to fasting, and to new ways of outward-bound interaction with the environment, like fight/flight, but also altering the environment to ease selection pressures (mouse holes for mice, etc.). Humans do not even internally manufacture all of their essential amino acids, relying on these open interaction systems to obtain them from their environment. (Which means that any constraint closures required for organism autonomy must be understood relatively to what can be regulated through external interaction and not only internal metabolic activity.) Just as with flight, all this also transforms ecological organisation.

In sum, if I might exploit a flight metaphor, when it comes to the expansion of life on the planet, it may be evolutionary selection that provides the thrust, but it is organisation that provides the lift. It is, as Howard Pattee taught us, the coordination of organisational constraints that makes possible the accumulating diversity and complexity of life. If organisation without evolution is impotent, evolution without organisation is blind.

Integration of Science and Philosophy

The dominant tradition in (meta-)philosophy is that philosophy and science are not to be integrated because philosophy provides an a priori normative framework for the analysis, conduct and evaluation of science whereas science constructs a posteriori empirical knowledge of the world by applying that framework. But in practice the development of understanding has rarely (really: never) happened like this. Philosophers have always borrowed ideas, theories and methods from science, and vice versa, each fertilising the other, unregarding of the proprieties of doing so. This has been a GOOD THING for both parties, each informing the other and keeping it on its toes. A minority naturalist (meta-)philosophy position would also applaud this intercourse as entirely appropriate. And that is what our authors consciously practice. Here is what they say (see Introduction): "... the approach developed in this book lies in between philosophy and theoretical biology. It deals with philosophical questions, like the nature of autonomy, agency and cognition, as well as their relations with concepts such as function, norms, teleology and many others; yet, it addresses these questions in close connection to, or even deeply entangled with, current scientific research." What emerges from this rich process is a coherent, if unfinished, majestic view of life as a subtly mutually entangled, organised whole from molecules to macro-ecology.

On Chasing Hares

Like all really interesting books, this book is profoundly incomplete: it starts new hares (new lines of thought) running on almost every page. This leaves the curious and/or thoughtful reader to enjoy the pleasure of identifying them and deciding which ones to follow up. A fine example already occurs in Chap. 1, in the nature of the closure found in self-regeneration and its relation to dynamical constraints. This issue is central, for according to the book's story there is no function or organisation, properly so-called, without closure ("an organization is by definition closed and functional", Chap. 3) and hence no autonomy either. I have previously mentioned constraints five times, including in characterising autonomy itself, and closure thrice, as if both notions were well understood. Did you notice any hares leap?

Closure has been an issue in thinking about autonomous systems from the beginning (see the summary in their Chap. 1). Founders like Varela emphasised closure as the distinctive feature of biological organisation and made its discovery at multicellular levels the key requirement for understanding them, even though closure was hard to uncover (it was thought to characterise the immune and nervous systems) and seemed to pull against the increasing interactive and organisational openness that marks multicellularity (see above). Many (myself included) adopted a process model: processes are sequences of dynamical states and process closure occurs where these states cycle through a closed loop of states, returning each time to an initial state, e.g. the normal or "resting" metabolism state. The cellular Krebs cycle is again a useful example. The thermodynamic flow, another process, drives the cycling, thus reconciling openness (flow) with closure (cycling). But Moreno and Mossio find this unsatisfying (for reasons I leave to the reader to pursue) and have developed their own distinctive account on which it is constraints that are closed and not processes, which are open on account of the thermodynamic flow. By constraint closure is meant, roughly, that the constraints so interrelate as to reconstitute one another. (So there is still a process cycle, but it is among constraint conditions, leaving thermodynamic processes to remain open.) To make the distinction between constraints and processes really sharp, they require that constraints do not interact, in the sense of exchange energy/materials, with the thermodynamic flow, only shape its direction. Think of a river flowing between frictionless banks. For this reason, they characterise constraints as not being thermodynamic entities and in Chap. 2 they support that by arguing that they are emergent entities with respect to the thermodynamic flow.

What are constraints, these non-thermodynamic entities that somehow shape the flow while not being of it? In standard mathematical dynamics constraints appear in the application of dynamical models where, although not directly represented in the system dynamics flow equations, they apply forces that constrain the dynamical possibilities of the flow. When they do not interact with the flow (ironically for Moreno/Mossio) those forces can be calculated and, like all modelled forces, are grounded in physical configurations of matter and/or fields of the same sort as make up the system being modelled, just located externally to it. But in autonomous systems all the matter/fields that give rise to those constraint forces have themselves to be assembled within the autonomous system itself in consequence of its constrained flow. Precisely that is the trick of autonomous self-regeneration, and a problem for understanding constraints.

For this means that constraints repeatedly degrade and have to be physically reconstructed, waste molecules literally replaced with new ones, etc. That is, the system itself must do work on its own constraints, or anyway on the matter/fields that give rise to them. Think of a real river that erodes and reconstructs its own banks as it flows. But that raises a first important issue: we have no workable methods for formulating the dynamics of systems that do work on their own constraints, the standard techniques of Lagrangian dynamics break down in this case. (See my "On the import of constraints in complex dynamical systems", *Foundations of Physics*, 2013, and earlier in Hooker (2011) above.) So how exactly are we to understand these systems and their self-reconstituting constraints? (Hare 1) This issue applies much more widely than biology, of course, since the self-formation and transformation of internal constraints is a major feature of complex dynamics anywhere (Hare 2). And, as noted above, but not in the book, apparently multicellular closed constraints have to be understood relative to an organism's interactional (agency) capacities, which itself depends on its functional, so closure, organisation (Hare 3).

And the manner in which Moreno/Mossio move to avoid facing the problem for autonomous dynamics (by requiring that constraints do no work and have none done on them) raises a second important issue: since constraints have to be reconstituted there are presumably periods of time when work is being done on at least some of them (on their supports): what kind of dynamics then applies to them and the flow? (Hare 4) These concerns are reinforced by a vivid picture in Chap. 3 of self-maintenance extended over time, for both intra-organism and inter-generational autonomous organisations, reinforced by the argument in Chap. 6 that developmental processes are necessary to multicellular constitution. (There is another group of hares loitering around these ideas.) But perhaps it also offers a way out in its conception of transmission of causal organisation over time that does not seem to require continuous satisfaction of closure (Hare 5). Even then, the hare 4 issue would remain to be addressed. And a further issue arises: considering time periods during which various proportions of constraints do not exist as such because they are doing work on some part of the system (including regenerating other constraints) and/or having work done on them (being regenerated), how large can those time spans be before system autonomy is considered disrupted and no longer explanatory of that system, and why? (Hare 6)

No doubt the authors will have anticipated such issues and been thinking about responses. (Their remarks on river banks and in a few other places reflect my earlier probings.) Irrespective, these questions should not be considered criticism of the book; to the contrary, they represent questions that could not be asked until the refined treatment of constraint closure Moreno and Mossio propose was available. And while there are lots of hares to startle, as there must inevitably be given our ignorance and a penetrating book, the present book succeeds in blunting many of the criticisms (including mine) made of the organisational approach. For instance, theirs is a position that takes the nature and role of biological organisation far beyond simple self-organisation of the kind beloved of the complex-behaviour-from-simple-rules-among-many-components tradition in the physics of complex systems. Indeed, that latter kind of process includes forming crystal lattices and like, so in fact it has no direct relationship at all with the kind of nested-complementary-correlations-and-regulations-among-disparate-components that this book is concerned with. The former could in principle be extended to encompass biology via bringing all organic chemistry under atomic modelling, but even then "organisation" in "self-organisation" remains a misnomer. (Two more hares.) Again, the book's position takes external interaction (individual and evolutionary) as seriously as internal organisation, whereas there are other traditions (discussed in the book) that are more closed-off to its importance, e.g. as illustrated above for understanding multicellular capacities. Nonetheless, we may still wonder whether the full extent of the interactive openness has been appreciated: what would their account of consuming internal organs under stress or adapting closure to environmental extraction of amino acids look like? (Another hare.) Finally, here the organisational approach is used to illuminate a thoroughly embodied approach to mind, for example with a deep connection developed to body plan, that counters the concern with "lifting off" an abstracted organisational pattern that has only nebulous connection to nervous system dynamics, organisation and functioning. However, there is still room to wonder about how neural phenomena characteristic of neural networks, whether distributed representations or waves, fit with organisation. (Another hare.) In these and like ways, this book represents a marked step forward in developing the organisational approach.

Meanwhile, there is the serious fun of chasing down such interesting and epistemically rewarding hares.

Conclusion

The authors describe my review of the draft of this book as, among other things, "relentlessly critical" (see closing remarks, Introduction). This is a compliment to both parties. A decade or more earlier I had entertained the prospect of a book on autonomy and discussed the idea with Moreno – on one occasion after an ocean swim near my Australian home and over a little local sauvignon blanc with freshly shucked Sydney Rock oysters, which he commented were "the best oysters I have ever tasted". (The preceding year at his coastal village I ate the best turbot I had ever tasted.) I hopefully suggested that the book could begin by understanding life through a series of ever tightening dynamical and thermodynamical constraints culminating with a notion of autonomy as the unique allowed evolvable organisation, just as the Krebs Cycle is a solution to capturing free energy for the cell. "Go ahead!" he said, "Be quick! I shall eagerly await your analysis." Of course, he knew better from years of trying just how hard that scientific task would be, still impossibly hard today where, for example, simple chemical cell models are still under development. I should have paid more attention to the quiet twinkle in his eye.

But we can all pay attention to what has been achieved. This book has thrust and lift. It is a masterly account of the organisational foundations of life, a splendid flight in the firmament of conception and understanding.

Professor Emeritus of Philosophy Cliff Hooker
Fellow of the Australian Academy of Humanities
PhD (Physics, Sydney University)
PhD (Philosophy, York University, Canada)

Contents

Introduction

Life as Autonomy

If we were to point out in a few words what characterises the phenomenon of life, we would probably mention the amazing plasticity and robustness of living systems, the innumerable ways they adapt, and their capacity to recover from adverse conditions. All these capacities have been on the surface of our planet since the origins of life, and for this reason we have become accustomed to seeing life as something almost "normal". And yet, looking at it from a more global perspective, life is quite an extraordinary phenomenon. In a short period of time (compared to the history of the universe), in a very tiny portion of the cosmos, a set of entities has managed to attain extremely improbable configurations, to keep them in far-from-equilibrium conditions, and to thrive under these conditions: self-organising, proliferating, diversifying, and even increasing their complexity. Furthermore, this persistently organised system (or, rather, this global system formed by millions of local, individualised systems, which combine decay and reproduction) has been able to deploy a set of selective forces, modifying its environment so as to enhance its own maintenance. In a word, life seems to be at the same time an extraordinarily precarious (and improbable) phenomenon and a powerful, robust, and easily expansive one.

Actually, this astonishing capacity to maintain highly organised systems seems to be the easiest way to recognise universally living matter beyond the specificities of terrestrial life. Present-day theories estimate that the universe came into being 13.7 billion years ago, while our planet was formed approximately 9 billion years later. In this period of time, or perhaps later, forms of organisation similar to early living systems on our planet possibly appeared in other parts of the universe. Indeed, if life appeared on our planet when certain physicochemical conditions were met, other planets with similar conditions could also have once supported forms of life. This raises the question of how we could recognise these hypothetical extra-terrestrial living systems, and what would be the essential features of *any* form of (possible) life. In the last decades, this question has been widely discussed.

For some (Cleland and Chyba 2007), it is impossible to say how such "essential features of life" should be conceived, because we only know life as it manifests itself on Earth. Yet, if what we mean by "life" is any material organisation that has evolved from non-living physicochemical systems (therefore obeying the universal laws of physics and chemistry) and has attained at least a degree of complexity capable of generating the properties we associate with the simplest forms of terrestrial life, we should be capable of recognising it anywhere in our universe, regardless of how differently these systems may be constituted (Ruiz-Mirazo et al. 2004). At the same time, the huge variety of life forms that have appeared during the very long history of life on our planet (Ward and Brownlee 2004) might downplay the argument that we have had access only to a unique example of life among a hypothetically huge set of extra-terrestrial biological systems. Be that as it may, when facing the question of the nature of life, we could not do otherwise than formulate theories based on – and tested against – life as we know it.

It is because of its capacity to achieve and maintain higher degrees of complexity that physical sciences find it very difficult to explain how life has originated. For this reason, the question of the origin of life is deeply entangled with the question of its very nature. Is there some law or principle in the physical world that allows explaining the emergence of life as a necessity or, as Monod (1970) thought, is the origin of life so unlikely that it is almost a miracle? How could inert matter originate something that seems to be so deeply different in its properties?

From the perspective of the physical sciences, explaining life is a highly challenging task because the more complex a system is, the less probable it becomes both in its appearance and its persistence. At first approximation, it might be easy to understand how simple building blocks may spontaneously generate composite stable structures (atoms, molecules, macromolecules . . .) due to different levels of forces (Simon 1969): as a result of these interactions, increasingly complex stable structures appear (endowed, in many cases, with new interactive properties, not present in their separate parts, such as superconductivity, chemical affinity . . .). As the complexity of the structures increases, however, its maintenance becomes a problem: thermal noise increases fragility and, moreover, the coincidence or coordination of many highly specific processes becomes increasingly unlikely.

It is true that recent advances in thermodynamics explain the formation of composite aggregates (called "dissipative structures"), whose parts are tied together without intrinsic forces, ensuring their cohesion in far-from-equilibrium conditions. However, as we will discuss at length in this book, these systems appear spontaneously and persist only when specific external boundary conditions are met and, more importantly, they lack internal complexity and functionality. In contrast, biology deals with highly complex systems, so that something more than initial conditions and fundamental laws seems to be required to explain a world of complex biological systems.

Assuming that nature does not make leaps and that, therefore, there is a continuum between non-living matter and life,[1] there should be explanatory principles of the transition from non-living to living matter. As Fry (2000) has pointed out, the fundamental problem of the origin of life lies in the tension between the principle of continuity and the difficulty of explaining the obvious differences between non-living and living matter. If the origin of life is a legitimate scientific question (and we think it is), one should look for a theory that bridges the gap between physics and biology. In particular, since living beings are made of the same constituents as non-living entities, what is the nature of the organisation that enables them to achieve, maintain, and propagate such a high degree of complexity? And what are the consequences of this extraordinary capacity?

On our planet, life has developed for a long period of time and has colonised the most diverse environments – from the deep oceans or even several kilometres under the Earth's crust to the upper levels of the atmosphere; from the hottest environments (over 100°C) to extremely acid or radioactive ones. And if we consider life from an historical perspective, it is even more impressive how it has managed to adapt to the successive catastrophic events that have occurred on our planet during the last 3.5 billion years. Admittedly, only the simplest forms of life are capable of such extreme robustness and versatility; at the same time, these forms of life have also been able to innovate and evolve towards increasingly higher levels of complexity. Life, as it has developed on our planet, has gradually integrated more and more levels of organisation (from unicellular life to colonies, multicellular organisms and societies).[2]

How can we explain all this diversity and complexity? Ever since Dobzhansky's (1973) famous dictum that "nothing in Biology makes sense except in the light of evolution", mainstream thinking in biology has seen evolution by natural selection as the source of diversity at every level of biological organisation. Indeed, the unfolding of an evolutionary process by natural selection, based on heritable genetic mechanisms, allows life to explore many possible combinations and solutions in order to survive. And the evolution-centred view of life has been so dominant that the idea of organism (which played a key role in nineteenth century biology) has become almost dispensable (Morange 2003). However, in a very fundamental sense, we shall argue at length that the reality is rather the opposite: evolutionary mechanisms operate because they are embodied in the complex organisation of organisms. Thus, if we look for the roots of the impressive capacity of life to proliferate, to

[1]Philosophically, this assumption amounts to adopting a monistic stance. Chapter. 2 is devoted to a detailed analysis of the position of the autonomous perspective developed in this book in the debate on emergence, reduction, and related issues.

[2]Nowadays we know that this process of diversification and complexification is not a contingent fact, but rather something "inscribed" in the evolutionary nature of life. As Gould (1994) has argued, evolution is not aimed towards an increase in complexity; in fact, life originates in the simplest form and many organisms have remained successfully as such. However, a few organisms occasionally introduced innovations, "thus extending the right tail in the distribution of complexity. Many always move to the left, but they are absorbed within space already occupied".

create an enormous variety of forms, to adapt to completely different environments, and particularly, to increase its complexity, we shall focus on individual living entities, namely on organisms, because evolution[3] as an explanatory mechanism actually presupposes the existence of organisms. As Varela (1979) pointed out,

> evolutionary thought, through its emphasis on diversity, reproduction, and the species in order to explain the dynamics of change, has obscured the necessity of looking at the autonomous nature of living units for the understanding of biological phenomenology. Also I think that the maintenance of identity and the invariance of defining relations in the living unities are at the base of all possible ontogenetic and evolutionary transformation in biological systems (p. 5).

As Rosen also emphasised, the crucial question for understanding life lies in the nature of its organisation.[4] It is true that any known living being cannot have appeared except as a result of a long history of reproductive events, since such a complex organisation can only be originated through an accumulative historical process and, furthermore, that its long-term sustainability also requires inter-generational entailments. This is clearly reflected in the fact that, in order to be operational, genetic components (which contribute to specify the metabolic machinery and organisation of single biological entities) must be shaped through a process that involves a large number of individual systems and many consecutive generations, or reproductive steps. Yet, this does not mean that the organisation of organisms should be neglected; on the contrary, a theory of living organisation is fundamental for understanding how these evolutionary mechanisms could have appeared and how they could work.

A theory of the living based on the concept of organism aims to review the concept of evolution and its role in a new way, attempting to overcome the dichotomy – and often opposition – between what since Mayr's (1961) work is called the biology of proximate causes and that of ultimate causes. Our vindication of the central role played by the notion of organism in biology should be placed within this wider perspective, in which the explanatory emphasis is placed on organisation. As Hooker and Christensen (1999) have highlighted, in order to

[3]The term 'evolution' could be understood in a very broad sense, just as an historical process of causal entailments. However, since Darwin, the term evolution has acquired a more restrictive sense, referring to specific mechanisms of inheritance and several other conditions (see for example, Godfrey-Smith (2009)). We will discuss the relation between autonomy and evolution in Chap. 5; here, we use the more restrictive sense of the term.

[4]"We cannot answer the question (...) 'Why is a machine alive?' with the answer 'Because its ancestors were alive'. Pedigrees, lineages, genealogies, and the like, are quite irrelevant to the basic question. Ever more insistently over the past century, and never more so than today, we hear the argument that biology *is* evolution; that living systems instantiate evolutionary processes rather than life; and ironically, that these processes are devoid of entailment, immune to natural law, and hence outside of science completely. To me it is easy to conceive of life, and hence biology, without evolution" (Rosen 1991: 254–55).

properly understand the evolution of biological systems, traditional approaches need to be embedded within a more general dynamical-organisational theory.[5]

Therefore, it is at the level of organisms, understood as cohesive and spatially bounded entities, that the biological domain's organised complexity is fundamentally expressed. Seen from the perspective of their relations with their environment, individual organisms are systems capable of acting for their own benefit, of constituting an identity that distinguishes them from their environment (at the same time as they continue interacting with it as open, far-from-equilibrium systems). This capacity of living beings to act for their own benefit follows from their peculiar form of organisation.

Living beings are systems continuously producing their own chemical components, and with these components they build their organs and functional parts. In a word, their organisation is maintaining itself. This is why living systems cannot stop their activity: they intrinsically tend to work or they disintegrate. Actually, this inherent tendency of living entities to promote their own existence – to act on their own behalf – could be related to the idea of the *conatus*, to which Spinoza (1677/2002) refers to designate the innate inclination of any entity to continue to exist and enhance itself.[6]

The root of this drive to persist lies in the principles of biological organisation. As Jonas (1966/2001) pointed out, the organisation of living systems is characterised by the inseparability between what they are – their "being" – and what they do – their "doing". This feature is reflected in their metabolism, which consists of a set of processes that allow them to build and replace their structures, grow and reproduce, and respond to their environments. Metabolism is the ongoing activity by which living beings continuously self-produce (and eventually, re-produce), self-repair, and maintain themselves. Unlike the Cartesian argument (which has had so much influence during modernity[7]) that living beings are like man-made machines, Kant was the first author who defended the view that organisms are

[5]As a matter of fact, an organisational perspective seems to be taking shape in the new evolutionary developmental biology, which studies how the dynamics of development determine the phenotypic variation arising from genetic variation and how this affects phenotypic evolution (Laubichler and Maienschein 2007).

[6]As Spinoza (1677/2002) writes, "Each thing, insofar as it is in itself, endeavors to persist in its own being" (Ethics, part 3, prop. 6). This is understood as an intrinsic tendency or force to continue to exist. Striving to persevere is not merely something that a thing does in addition to other activities it might happen to undertake. Rather, striving is "nothing but the actual essence of the thing itself" (Ethics, part 3, prop. 7). See Duchesneau (1974) for an in-depth analysis of Spinoza's account of living systems, and a comparison with the Cartesian one.

[7]Actually, the Cartesian distinction between *res extensa* and *res cogitans*, which subsumed the biological domain within a global mechanistic vision of nature, facilitated a scientific research programme for studying living systems. It should be underscored that, while Descartes' metaphysical dualism is widely recognised and is a prominent feature of his *Meditations*, scholars in the past generation have also focused on the complexity of his natural philosophy, including his work in physiology, medicine but also on the passions, as displaying something very different: a more 'integrated' view of bodily function. See notably the essays collected in Gaukroger et al. (2000).

deeply different from machines because their parts and activities are non-separable, and the functions of these parts are not externally imposed, but rather intrinsically determined. According to Kant (1790/1987), since the activity performed by the parts of the organism is carried out for their own maintenance, organisms are intrinsically teleological. As he writes in the *Critique of Judgement*:

> In such a product of nature each part, at the same time as it exists throughout all the others, is thought as existing with respect to the other parts and the whole, namely as instrument (organ). That is nevertheless not enough (because it could be merely an instrument of art, and represented as possible only as a purpose in general); the part is thought of as an organ producing the other parts (and consequently each part as producing the others reciprocally). Namely, the part cannot be any instrument of art, but only an instrument of nature, which provides the matter to all instruments (and even to those of art). It is then – and for this sole reason – that such a product, as organized and organizing itself, can be called a natural purpose (CJ, § 65).

This view allows him to open up a gap in the physical world, since organisms cannot be brought under the rules that apply to all other physical entities. Thus, Kant asks himself:

> How purposes that are not ours, and that we also cannot attribute to nature (since we do not assume nature to be an intelligent being) yet are to constitute, or could constitute, a special kind of causality, or at least a quite distinct lawfulness of nature (CJ, § 61).

This "special" kind of causality is circular, namely, effects derive from the causes but, at the same time, generate them. The very organisation of living beings, in which the parts generate the whole, and, conversely, the whole produces and maintains the parts, shows a kind of intrinsic purpose. Kant grounds the idea of purposiveness (and teleology) in the holistic and circular organisation of biological organisms and, more precisely, in the fact that they are able to organise by themselves, to *self*-organise.[8] Unlike artefacts, organisms are "natural purposes": they are not produced or maintained by an external cause, but instead have the self-(re)producing and self-maintaining character that is revealed in the kinds of vital properties they display (reciprocal dependence of parts, capacity for self-repair and self-(re)production).

Today, some aspects of the Kantian perspective are undergoing resurgence. For example, the recent blossoming of systems biology (Kitano 2002; Science, *special issue* 2002; Bogeerd et al. 2007), focused on the complexity of biomolecular interaction networks, is much closer to a holistic or integrative conception of living systems than the reductionist views predominant in molecular biology. Thanks to the development of new scientific tools, these more holistic theories place the question of the organisation at the centre of biological research. This recent trend contrasts with the preceding history of biology, during which the Kantian view has often be seen as marginal (even through this view has been corrected by

---

[8] Actually, Kant has been one of first authors to use the term "self-organisation". In Chap. 1, we will briefly mention how the meaning of this concept has progressively shifted during the 20th century.

the recent historiography, see for instance Huneman 2007; Richards 2000; Sloan 2002), essentially because it was thought to be at odds with the model of causality predominant in Newtonian science.

And yet, the Kantian perspective had continuity in the (mostly Continental) Biology of the nineteenth century, especially in the work of Goethe and Cuvier (Huneman 2006). In the first part of the twentieth century, many biologists were still convinced that the nature of living organisation – understood, following Kant's inspiration, as the form in which the parts interact with each other to bring forth the properties of the whole – was one of the main issues of biology. This view was commonly labelled *organicism* (Wolfe 2010; see also Gilbert and Sarkar 2000). Organicism considers the observable structures of life, its overall organisation, and the properties and characteristics of its parts to be the result of the reciprocal interplay among all its components. The organicist tradition was influential in early twentieth century biology. During the twenties and thirties, a group of researchers, including Woodger, Needham, Waddington, and Wrinch, created the "Theoretical Biology Club", whose objective was precisely to promote the organicist approach to biology. This movement – in which we can include other authors, like Bernal and Bertalanffy – was characterised by a predominant anti-reductionist and holistic inspiration (Etxeberria and Umerez 2006). Among these researchers, the name of Waddington is worth stressing because his work, after the Second World War, permitted the connection between the organicist movement of the thirties and the new tendencies of the sixties and seventies.

To understand the roots of the current blossoming of the "Kantian-inspired organicist ideas" in biology during the twentieth century, let us mention some other scientific trends, falling outside the frontiers of biology.

First, during the thirties and forties, a number of physicists associated with the development of quantum theory, interested in the nature of biological organisation, turned their attention to biology. Among these scientists, it is worth emphasising the name of Schrödinger, who gave his famous lectures "What Is Life?" in 1943 (Schrödinger 1944). Following this work, other quantum physicists addressed the problem of what characterises the specificity of living systems with regard to physical ones. Among these we can include researchers like von Neumann and Pauli. Interestingly, the advances in physics inspired new attempts to challenge reductionist assumptions. For example, Rashevsky, according to his disciple Rosen, defended

> a principle that governs the way in which physical phenomena are organized, a principle that governs the organization of phenomena, rather than the phenomena themselves. Indeed, organization is precisely what relational biology is about (Rosen 1991: 113).

During the seventies, Rosen himself and Pattee (Umerez 2001) also developed an anti-reductionist view of the specific organisation of living systems, based on his analyses of the specific causation associated with emergent constraints that living systems generate (see further below).

Second, special emphasis should be put on the cybernetic movement. The cyberneticists were influenced by the work of the American physiologist Cannon

(1929) who, in the early 1930s, developed the concept of "homeostasis" (whose origins date back to the work of the French biologist Claude Bernard[9]) as a key feature of the organisation of living beings. According to Cannon, the idea of homeostasis expresses the tendency of living systems to actively maintain their identity, despite external perturbations or differences within their environment. During the 1970s, a new generation of cyberneticists, notably Von Förster, Ashby, and Maturana, created the so-called second-order cybernetics. This movement was especially interested in the study and mathematical modelling of biological systems, based on the ideas of recursivity and closure (Cahiers du CREA 1985). Second-order cybernetics is of special relevance for our purposes, since it constituted the scientific environment in which the theory of *autopoiesis* was elaborated (see below).

Third, after the work of Prigogine (1962), the idea of self-organisation in far-from-equilibrium conditions began to enter into scientific discourse in physics, which also helped the Kantian view to gain influence in biology. Yet, as we will discuss at length in Chap. 1, there is an important conceptual difference between the Prigoginian concept of (physical) self-organisation and the Kantian notion of (biological) organisation. As Fox Keller (2007) has pointed out, the kind of complexity of organisms resulting from an iterative processes of organisation that occur over time is completely different from the one-shot, order-for-free kind of self-organisation associated with some kind of non-linear dynamical systems. In particular, the former is constituted by functional parts, whereas the latter lacks functionality. The logic of the metabolism, for example, shows a functionally diversified organisation, clearly different in this sense from any physicochemical dissipative structure. In this sense, as we will see, what we need is a view of biological systems that goes beyond a generic vindication of an organisational-centred biology. What matters is the understanding of the *specificity* of the organisation of biological systems, which are not just self-organised systems.

In the second post-war period, both the New Synthesis in evolutionary biology and the revolution of Molecular Biology created a scientific atmosphere that was quite unprepared to accept organicist and Kantian views (Moreno et al. 2008). Accordingly, this tradition remained, until very recently, marginal in biology. In this context, however, Waddington was the main driver of a movement that advocated an organisational approach in biology, by reviving the "first" Theoretical Biology of the twenties and thirties (Etxeberria and Umerez 2006). This "second" Theoretical Biology was initially developed by several pioneering authors like Waddington himself (1968–1972), Rosen (1971, 1972, 1973, 1991), Piaget (1967), Maturana and Varela (1980), Pattee (1972, 1973), and Ganti (1973/2003, 1975). Many of these authors put strong emphasis on the idea that the constitutive organisation of biological systems realises a distinctive regime of causation, able not only of producing and maintaining the parts that contribute to the functioning of the system as an integrated, operational, and topologically distinct whole but also able

[9]See Bernard (1865) and (1878).

to promote the conditions of its own existence through its interaction with the environment. This is essentially what we call in this book *biological autonomy*.

To give a preliminary idea of what autonomy is about, let us mention one of its first and well-known accounts, the theory of *autopoiesis* proposed by the Chilean biologists Maturana and Varela in the early 1970s (Maturana and Varela 1973; Varela et al. 1974). In the theory of autopoiesis, although the concept of autonomy is applied to different specific biological domains (immune, neural . . . see Varela 1979), it characterises the fundamental feature of the living, namely, the autopoietic organisation. Autopoiesis refers to the capacity of self-production of biological metabolism, by emphasising (in a simplified and abstract way) its causal circularity – which Maturana and Varela called "operational closure". In particular, their model describes the production of a physical boundary, which is conceived as a condition of possibility of the internal chemical network (because it ensures suitable concentrations for the maintenance of the component production network); in turn, the network maintains the physical boundary (because it is the component production network which produces the special self-assembling components that build the membrane). In their own terms (in which the cybernetic flavour is manifest):

> An autopoietic machine is a machine organized (defined as a unity) as a network of processes of production (transformation and destruction) of components which: (i) through their interactions and transformations continuously regenerate and realize the network of processes (relations) that produced them; and (ii) constitute it (the machine) as a concrete unity in space in which they (the components) exist by specifying the topological domain of its realization as such a network (Maturana and Varela 1980: 78).

Thus, autopoiesis consists in a recursive process of component production that builds up its own physical border. The global network of component relations establishes self-maintaining dynamics, which bring about the constitution of the system as an operational unit. In short, physical border and metabolic processes are entwined in a cyclic, recursive production network and they together constitute the identity of the system. From this perspective, phenomena like tornadoes, whirlpools, and candle flames, which are to a certain degree self-organising and self-maintaining systems, are not autonomous, because they lack an internally produced physical boundary, and are not concrete topological units. In that sense, what distinguishes self-organisation from autonomy is that the former lacks an internal organisation complex enough to be recruited for deploying selective actions capable of actively ensuring the system's maintenance.

For the purposes of this book, it is worth mentioning two lines of criticism that have been addressed to the theory of autopoiesis. On the one hand, autopoiesis conceives autonomy as a fundamental internal determination, defined by the operational closure between the production network and the physical border. In this model, interactions with the environment do not enter into the definition-constitution of the autonomous system; rather, the interactions with the environment – that Maturana and Varela called "structural couplings" – follow on from the specific internal identity of each autopoietic system. On the other hand, Maturana and Varela define autonomy in rather abstract and functionalist terms: material and energetic aspects are considered as purely contingent to its realisation.

On both these issues, the framework that we will develop in this book takes a different path. The autonomous perspective, we hold, should take into account the "situatedness" of biological systems in their environment, as well as their "grounding" in thermodynamics. As a matter of fact, these issues have been at the centre of the most recent studies on biological autonomy, by authors like Hooker, Collier, Christensen, Bickhard, Kauffman, Juarrero, and the IAS Research Group,[10] who have stressed that the interactive dimensions of autonomous systems in fact *derive* from the fact that they are thermodynamically open systems, in far-from-equilibrium conditions. As these authors have explained, since the constitutive organisation of biological systems exists only in far-from-equilibrium conditions, they must preserve an adequate interchange of matter and energy with their environment or they would disintegrate. For example, in Kauffman's approach, the main condition required for considering a system autonomous is that it should be capable of performing what he calls "work-constraint cycles" (Kauffman 2000). As Maturana and Varela, Kauffman's account envisages how autonomy can come out of the causal circularity of the system; yet, in his view, this circularity is understood not just in terms of abstract relations of component production but in explicit connection with the thermodynamic requirements that the system must meet to maintain itself.

In accordance with this literature, we will make in this book a conceptual distinction between two interrelated, and yet conceptually distinct, dimensions of biological autonomy: the *constitutive* one, which largely determines the identity of the system; and the *interactive* one, which, far from being a mere side effect of the constitutive dimension, deals with the inherent functional interactions that the organisms must maintain with the environment (Moreno et al. 2008). These two dimensions are intimately related and equally necessary. It might be illuminating to think of the example of the active transport of ions across the cell membrane, required to prevent osmotic crises. The cell can be maintained as long as ion transport is performed, but this interaction can only be carried out because there is a constitutive chemical organisation providing the membranous machinery that does the work. In particular, the emphasis on the interactive dimension implies, as we will stress repeatedly, that autonomy should not be confused with independence: an autonomous system must interact with its environment in order to maintain its organisation[11] (Ruiz-Mirazo and Moreno 2004). As we will discuss in Chap. 4, this is what grounds the agential dimension of autonomy.

[10]The IAS Research Group – to which the authors of this book belong – has been working since the last 25 years on autonomous perspective in biology, while extending it to other fields as cognition, society and bioethics. See also the end of this Introduction and footnote 6 in Chap. 1.

[11]Hooker has recently defined autonomy as "the coordination of the internal metabolic interaction cycle and the external environmental interaction cycle so as the latter delivers energy and material components to the organism in a usable form and at the times and locations the former requires to complete its regeneration cycles, including regeneration of the autonomy capacity" Hooker (2013).

Again, there is a reciprocal dependence between what defines the conditions of existence of the system and the actions derived from its existence: from the autonomous perspective, in Jonas's terms, the system's *doing* and its *being* are two sides of the same coin (see also Moreno et al. 2008). In this view, the environment becomes a world full of significance: facts that from the outside may appear just as purely physical or chemical develop into positive, negative, or neutral influences on the system, depending on whether they contribute to, hinder or have no effect on the maintenance of its dynamic identity. Even the simplest living organism creates a set of preferential partitions of the world, converting interactions with their surrounding media into elementary values, as we will explain extensively in Chaps. 4 and 7. Von Uexküll (1982/1940) called this subjective meaningful world of each organism *Umwelt*. The interactive dimension of autonomy is where the nature of living systems as inventors of worlds with meaning becomes manifest (see also Hoffmeyer 1996). Indeed, this aspect was recognised by Weber and Varela (2002) who argued, following Jonas, that autonomy implies a meaningful relation with the environment.

The autonomous perspective that we develop here endeavours then to grasp the complexity of biological phenomena, by adequately accounting for their various dimensions, specificities, and relations with the physical and chemical domains. As we will discuss throughout the book, our framework differs in many ways from preceding related models, mainly because we aim at – simultaneously – enriching and specifying their central tenets, in close contact with current scientific theories. In the remainder of this introduction, let us give a synthetic overview of the ideas that we will be advocating.

First, the self-determination of the constitutive organisation remains the conceptual core of autonomy. We share with existing accounts of autonomy the idea that biological systems are constituted by a network of causal interactions that continuously re-establish their identity (see also Bechtel 2007). The aim of Chap. 1 will be to provide an explicit conceptual and (preliminarily) formal account of self-determination in terms of what we will label "closure of constraints".

Biological systems determine (at least in part) themselves, we will contend, by constraining themselves: they generate and maintain a set of structures acting as constraints which, by harnessing and channelling the processes and reactions occurring in the system, contribute to sustain each other, and then the system itself. The core of biological organisation *is* the closure of constraints. We will discuss how the concept of closure allows specifying what kind of "circularity" is at work in the biological domain, and how it fundamentally differs from other "process loops" and self-organising phenomena in Physics and Chemistry. In particular, we will emphasise that biological closure requires taking into account, at the same time, the conceptual distinction, and yet inherent interdependence, between two causal regimes: the constraints themselves and the thermodynamic flow on which they act. In the autonomous perspective, closure (of constraints) and (thermodynamic) openness go hand in hand. Self-determination as closure constitutes the pivotal idea on which we will build our account of autonomy. A first step is made in the last section of Chap. 1, where we will claim that biological organisation, to be such, requires regulation. The long-term preservation of biological organisms supposes the capacity to self-maintain not only in stable conditions but also, and crucially, before potentially deleterious internal or external perturbations. In such circumstances, regulatory capacities govern the transition towards a new viable situation, be it by countering the perturbations or by establishing a new constitutive organisation. In all cases, we will account for regulation in terms of a specific set of constraints, which contribute to the maintenance of the organisation *only* when its closure is being disrupted: accordingly, we will argue that regulatory constraints should be understood as being subject to *second-order* closure.

Does the autonomous perspective require appealing to some form of emergentism? In previous years, some authors have argued that accounts dealing with concepts like self-organisation, closure, constraints, autonomy, and related ones are indeed committed to the idea that biological organisation is an emergent determination. In Chap. 2, we deal with this issue, advocate a monistic stance, and provide a twofold argument. First we argue, against exclusion arguments, that closure can be consistently (with respect to our monistic assumption) understood as an emergent regime of causation, in the specific sense that the relatedness among its constituents provides it with distinctive and irreducible properties and causal powers. Second, although the closure of the constitutive organisation makes sense of the claim that "the very existence of the parts depends on their being involved in the whole", we hold that closure does *not* imply inter-level causation, in the restrictive sense of a causal relation between the whole and its own parts (what we label "nested" causation). Yet, we leave room for appealing to nested causation in the biological domain, if relevant cases were identified and the adequate conceptual justification were provided.

With these clarifications in hand, Chap. 3 addresses the question of the distinctive emergent features of organisms by arguing, in particular, that the closure of constraints provides an adequate and naturalised grounding for the teleology, normativity, and functionality of biological organisation. When closure is realised, the existence of the organisation depends, as we have already emphasised, on the effects of its own activity: accordingly, biological systems are teleologically organised, in a specific and scientifically legitimate sense. Because of teleology, moreover, the activity of the organism has an "intrinsic relevance" which, we submit, generates the norms that the system is supposed to follow: the system must behave in a specific way, otherwise it would cease to exist. Hence biological organisation, because of closure, is inherently normative. And then, by grounding teleology and normativity, closure grounds also functionality in biological organisation: the causal effects produced by constraints subject to closure define biological functions. The general upshot of the analysis, at the end of Chap. 3, will be the deep theoretical binding between "closure", "organisation", and "functionality": it will be our contention that, from the autonomous perspective, they are reciprocally defining concepts, which refer to the very same causal regime.

The constitutive dimension of closure, however, is not autonomy. As mentioned in the preceding pages, autonomy also includes an interactive dimension, dealing with the relations between the organism and its environment. We deal with the interactive dimensions in Chap. 4, and refer to it as *agency*, characterised as a set of constraints subject to closure, exerting their causal effects on the boundary conditions of the whole system. At the end of the chapter, we argue that a system whose organisation realises closure, regulation, and agency, as defined in the first part of the book, is an autonomous system, and therefore a biological organism. More precisely, Chap. 4 elaborates on a definition of minimal autonomy that captures the essential features of biological organisation in its (relatively) most simple manifestations, typically in unicellular organisms.

What has the autonomous perspective to say about more complex organisations and specifically about multicellular organisms? One of the main weaknesses of the organisational tradition in biology is arguably the fact that it has never explicitly addressed the issue of higher *levels* of autonomy: How many levels of autonomy can be identified in the biological realm, and what are their mutual relations? In Chap. 6, we make a first step in this direction: we try to frame the issue of higher-level autonomy in precise terms and submit some explicit hypotheses on its features. The central idea will be that what matters for higher-level autonomy is *development*. More specifically, multicellular systems are relevant candidates as organisms when their organisation exerts a functional control over the development of unicellular components, so to induce their differentiation which, in turn, makes them apt to live only in the very specific environment constituted by the multicellular system: in a word, the control over development produces the relevant degree of *functional integration* that distinguishes multicellular organisms (as autonomous systems) from other kinds of multicellular systems. What about the relations between levels of autonomy? In spite of their differentiation (and then of the loss of some of their capacities), we will argue that unicellular constituents of higher-level organisms still meet the requirements of autonomy. In fact, the very possibility of higher-level autonomy seems to require that lower-level entities preserve an adequate degree of complexity: multicellular autonomy requires unicellular autonomy. One of the objectives of Chap. 6 (and partly of Chap. 4, last section) will be, by relying on an explicit definition of autonomy, to provide relevant criteria for examining different kinds of higher-level associations and organisations of autonomous systems and to compare them on theoretical grounds. In particular, our framework could allow locating them in a *continuum* of organised systems going from, at one extreme, those cases fulfilling only the requirements for closure (as ecosystems) to systems being progressively more integrated (as the cyanobacterium *Nostoc punctiforme*), up to genuine multicellular organisms (higher-level autonomous systems) at the other extreme.

The transition to multicellular autonomy paves the way towards cognition, which is possibly the most amazing innovation during the evolution of life. Cognition, as discussed in Chap. 7, is much more than a complex form of agency. It is better conceived as a radically new kind of autonomy whose specific features and dynamics go, qualitatively, far beyond multicellular autonomy, opening the way towards our own origins as human beings. In this sense, the analysis of cognition is related to the nuclear problem of the gap between the "biological" (broadly understood) and the "human" domains. Yet, the autonomous perspective strives to understand and explain cognitive capacities in close connection to a bodily organisation, which is in turn the product of a long evolutionary process, through which new phenomena such as emotions or consciousness – and a world of meaning and values – have been generated. The appearance of cognition is the result of the evolution towards increasingly higher degrees of both constitutive and interactive complexity: in this sense, with all its specificities, cognition is still a "biogenic" (Lyon 2006) phenomenon. By framing the issue of cognition in these terms, we think that it can be better handled in naturalised terms, without underestimating the formidable difficulties that any satisfactory account of cognition has to face to understand its complex nature and phenomenological novelty. Accordingly, Chap. 7 is possibly the most ambitious and yet incomplete, since it sketches in a preliminary way many problems for which much more work will be required.

Autonomy, as conceived in this book, lies at the intersection between different dimensions, and specifically the constitutive and interactive ones, on which we put strong emphasis in the previous pages. Yet, this is not the whole story: autonomy also has a *historical* dimension. As we will discuss in Chap. 5, no adequate understanding of the emergence of autonomous systems (and specifically highly complex autonomous systems, as present biological organisms) can be obtained without taking into account the evolutionary process that brought them about. Autonomous systems are too complex to be spontaneous and cannot self-organise (in the sense of generate themselves) as dissipative systems do: their complexity requires an evolutionary process of accumulation and preservation. Yet, in addition to acknowledging the fundamental place of history in the autonomous perspective, we will submit two related ideas. First, the historical dimension does not have the same theoretical status as the constitutive and interactive ones: while the latter two *define* autonomy, the former does not. The reason is that we do not need history to understand what biological systems are, but rather to understand where they come from: these two questions are of course related, but conceptually distinct. Second, we will restate the relations between selection and organisation, by advocating the general picture according to which the evolution of biological systems stems from the mutual interplay between organisation and selection: this is because, as we will argue, organisation is a condition, and not only an outcome, of evolutionary processes.

Having outlined the central ideas of the book, let us point out that it is, of course, not our intention to develop an exhaustive account of biological autonomy, which would deal with all aspects and implications of the philosophical and theoretical framework. Rather, our ambition is to offer a coherent and integrated picture of the autonomous perspective, by focusing on what we think are some of its central tenets. Much more could (and hopefully will) be written on biological autonomy, but we hope that the ideas of this book can be a useful ground on which future investigations will rely.

This book is the result of a collaboration that goes far beyond that between the two authors. After having promoted (together with Julio Fernandez, Arantza Exteberria, and Jon Umerez), more than 20 years ago, the creation of the *IAS Research Centre for Life, Mind and Society*, at the University of the Basque Country, in Donostia – San Sebastian, Alvaro Moreno has had since then the chance to work in this highly stimulating intellectual environment. In this respect, a special thought goes to Francisco Varela, who has been a fundamental source of inspiration for the creation of the *IAS Research* group and, for many years afterwards, a close collaborator and a friend.

Matteo Mossio joined the group in 2008 as a postdoctoral fellow and, after having moved back to Paris in 2011, maintains close collaborations with many of its members. Since the constitution of the *IAS Research* group its members have collectively developed the autonomous perspective in the biological, cognitive, biomedical, and ethical domain. The ideas developed in this book, then, are deeply grounded into the substantive and extensive philosophical and theoretical work undertaken by our colleagues and friends.[12]

It then goes without saying that we are intellectually indebted with many people. Let us thank first those who co-authored previous publications with (at least one of) us and allowed us to rework and use in this book some of the ideas advocated there: Argyris Arnellos, Xabier Barandiaran, Leonardo Bich, Maël Montévil, Kepa Ruiz-Mirazo, and Cristian Saborido. At the beginning of each chapter, we inserted a note in which we give the references of the specific publications from which some ideas and text portions have been taken and adapted.

We are sincerely grateful to the other members of the IAS Research Centre for continuous interactions, over the years, on a variety of topics related to this book: Antonio Casado da Rocha, Jesús Ibañez, Hanne de Jaegher, Asier Lasa, Ezequiel di Paolo, and Agustin Vicente. Also, we thank many other researchers with whom one of us (AM) has worked for a long time: Francisco Montero, Federico Morán, Juli Peretó, and more recently, Nei Nunes and Charbel El-Hani.

In Paris, the whole *Complexité et Information Morphologique Team* (CIM), at the Ecole Normale Supérieure, deserves a special mention. Some years ago, Giuseppe Longo created a small but very active interdisciplinary team, nourished by the talent of several young fellows: among them, let us thank with special emphasis Nicole Perret and Paul Villoutreix. We would also like to express our deepest gratitude to Giuseppe, a remarkably brilliant and profound scientist, for his wise guidance on – and unfailing support to – Matteo's academic and scientific trajectory. Recently, Matteo has been invited, together with some of the CIM members, to join the new *Theory of Organisms* research group at the Ecole Normale Supérieure, supervised by Ana Soto. We warmly thank her, as well as Carlos Sonnenschein and Paul-Antoine Miquel, for this unique opportunity to engage in stimulating and quality discussions and exchanges.

Since Matteo's appointment, the *Institut d'Histoire et Philosophie des Sciences et des Techniques* (IHPST) has constituted a privileged scientific environment and

[12]Over the years, the activites of the *IAS Research* group have received funding by both Basque and Spanish public institutions. This book, in particular, was supported by grant IT 590–13 of the Gobierno Vasco, and grant FFI2011-25665 of the Ministerio de Industria e Innovación.

provided him with ideal conditions of work. For that, we want to thank its director Jean Gayon, as well as all the members and colleagues who, in many cases, have become good friends.

Lastly, we are greatly indebted to those colleagues and friends who took the time to read and critically comment on early versions of the manuscript: Philippe Huneman, Johannes Martens, Arnaud Pocheville, and Charles Wolfe. In most cases, their observations and criticisms were decisive to highlight some of the weaknesses of the arguments and force us to improve their clarity and accuracy. In this respect, we owe a lot to Alicia Juarrero, who has not only made a number of precise and lucid comments on various ideas developed in the book but also crucially contributed to bringing the initial unstable language closer to correct English. We also want to warmly thank Juli Peretó for his help in the elaboration of many figures.

The final acknowledgements go to Cliff Hooker: his meticulous, lucid, and uncompromisingly critical reading of the entire manuscript has induced substantial changes (and, we hope, improvements!) in the formulation of the ideas, regarding both the form and the content. Last but not least, he has kindly written the best foreword we could expect.

Donostia – San Sebastian/Paris
October 7th 2014

1
Constraints and Organisational Closure

The first and most fundamental tenet of the autonomous perspective is the idea that the constitutive dimension of biological systems is inherently related to self-determination. As we recalled in the introduction, what constitutes biological systems is the fact that the effects of their activity and behaviour play a role in determining the system itself. As autonomous systems, biological systems "are (at least in part) what they do".

To a first approximation, all accounts of biological autonomy developed during recent decades share this idea, which most of them refer to using the technical term "closure". Despite the differences in existing formulations, the concept of closure aims to ground the intuition about self-determination in a biologically relevant and treatable way. In very general terms, it designates a feature of biological systems by virtue of which their constitutive components and operations depend on each other for their production and maintenance and, moreover, collectively contribute to determining the conditions under which the system itself can exist (Mossio 2013).

The term was first used in the biological domain by Varela in his *Principles of Biological Autonomy* (Varela 1979), and was later adopted by several other authors, including Howard Pattee, Robert Rosen, and Stuart Kauffman, in a similar or complementary sense. Varela's account constitutes here a relevant starting point, since it explicitly establishes a theoretical connection between closure and autonomy through the so-called "Closure Thesis", according to which "every autonomous system is operationally closed" (Varela 1979: 58). Although the thesis does not enunciate an equivalence – which means that, for Varela, closure does not *define* biological organisation (a point on which other authors, such as Rosen, would disagree) – it does put closure at the core of biological organisation, viewing it as a necessary and constitutive feature of autonomy.

Some of the ideas exposed in this chapter, as well as some parts of the text, are taken from (Montévil and Mossio 2015).

A. Moreno, M. Mossio, *Biological Autonomy*, History, Philosophy and Theory of the Life Sciences 12, DOI 10.1007/978-94-017-9837-2_1

Since its formulation, the Closure Thesis has indeed remained a common assumption in the philosophical and scientific tradition centred around biological autonomy, and the concept of closure has being increasingly developed in recent theoretical, computational, and experimental studies (Chandler and Van De Vijver 2000).

Yet, in spite of the current interest in closure as a key notion for understanding biological organisation, it should be noted that no consensus has yet been reached regarding a precise definition. Of course, definitions are not a goal in themselves, and the degree of accuracy which is required may depend on the role played in the general framework. In the case of closure, in our view, the lack of precision does indeed constitute an obstacle for the further development of the autonomous perspective, since closure is a fundamental pillar of the whole account, on which many (or even most) other aspects rely either directly or indirectly, as we will discuss at length in the following chapters.

In particular, the very status of closure as a causal regime with distinctive properties remains somehow controversial since, to date, no explicit account of the relations between closure and other kinds of causal regimes at work in physics and chemistry has been offered. This is a crucial issue since it might be possible that all accounts of biological organisation referring to closure could be reformulated in terms of other physicochemical causal regimes without any relevant information being lost. If this were the case, the concept itself, as well as all other notions relying on it, would have some heuristic value for biological research, but no explanatory role. Consider for instance the central feature of closure, i.e. the mutual dependence[1] between constituents and their collective capacity to self-determine. At first glance, indeed, mutual dependence seems to be by no means a distinctive feature of the biological domain. Let us mention an example that is frequently referred to in this kind of debate, namely, the Earth's hydrologic cycle.[2] Here, a set of water structures (e.g. clouds, rain, springs, rivers, seas, etc.) generate a cycle of causal relations in which each contributes to the maintenance of the whole, and is in turn maintained by the whole. Clouds generate rain, which (contributes to) generates a spring, which gives rise to a river, which (contributes to) generates a lake, which regenerates clouds, and so on. Is this a case of closure?

Arguably, a large number of physical and chemical systems could be described as generating some form of mutual dependence of this kind between their constitutive entities and processes. As a consequence, a coherent account of closure has to choose between two alternative options: either closure is to be conceived as a specific variant of other kinds of causal regimes encountered in physics and

[1] Strictly speaking, "mutual dependence" and "closure" are not synonymous. While the former is realised by any (sub)set of entities which depend on each other, the latter is realised by the set of *all* entities which are mutually dependent in a system. So for instance, the heart and the lungs realise mutual dependence among them, but only the whole set of organs of the organism realises (by hypothesis) closure.

[2] Another, more complex, example is the atmospheric reaction networks, which realise a closed loop of chemical reactions (Centler and Dittrich 2007).

chemistry, in which case the difference between physicochemical and biological systems, in this respect, would possibly be quantitative, but not qualitative; or, alternatively, it might be that closure is qualitatively irreducible to most kind of physical and chemical regimes and dependencies, and so specific to the biological domain.

The aim of this first chapter is to propose a theoretical and formal framework that characterises closure as a causal regime specifically at work in biological organisation. In particular, it will be our contention that biological systems can be shown to involve two distinct, although closely interdependent, regimes of causation: an *open* regime of thermodynamic processes and reactions, and a *closed* regime of dependence between components working as constraints.

1.1 Biological Determination as Self-Constraint

In the introduction to *Toward a Practice of Autonomous Systems* (Bourgine and Varela 1992), restate the Closure Thesis and clarify that they build on an algebraic notion, according to which

> a domain K has closure if all operations defined in it remain within the same domain. The operation of a system has therefore closure, if the results of its action remain within the system (Bourgine and Varela 1992: xii).

Applied to biological systems, closure is realised as what Varela labels *organisational* (or *operational*) closure,[3] which designates an organisation of processes such that:

> (1) the processes are related as a network, so that they recursively depend on each other in the generation and realisation of the processes themselves, and (2) they constitute the system as a unity recognisable in the space (domain) in which the processes exist (Varela 1979: 55).

Varela's account is perhaps the best-known and most influential one within the autonomous perspective. It has several qualities, particularly that of providing a general and abstract characterisation, which can be realised in nature by different kinds of biological systems and sub-systems.[4] In each case, the nature and kind of components and processes subject to closure are different, as is the kind of unity that they generate. In the specific case of the cell, as mentioned in the Introduction, closure takes the exemplary form of autopoiesis (Varela et al. 1974), which is

---

[3]It should be noted that, over the years, Varela himself has proposed slightly different definitions of operational closure. Also, more recent contributions have introduced a theoretical distinction between organisational and operational closure: whereas "organisational" closure indicates the abstract network of relations that defines the system as a unity, "operational" closure refers to the recurrent dynamics and processes of such a system (see Thompson 2007).

[4]According to Varela, three realisations of closure have been described: the cell, the immune system, and the nervous system (see Varela 1981: 18).

realised at the chemical and molecular level, and involves relations of material production among its constituents. In all cases, organisational closure constitutes the *fundamental invariant* of biological phenomena, in spite of the variability of its concrete realisations.

Despite its qualities, however, we would underscore what we take to be the fundamental weakness of Varela's account of closure. The characterisation described above refers to the processes as the relevant constituents of the system that, when organised in a network, must realise mutual dependence and closure. It seems only fair to point out that, for Varela, closure is understood as *closure of processes*. And here, in our view, is where the problem lies. Formulated in these terms, closure can in principle be used to describe not only the constitutive organisation of biological systems – which are by hypothesis the prototypical example of autonomous systems – but also a number of physical and chemical systems such as, for instance, the famous hydrologic cycle.

To this objection, one may reply that the definition emphasises the "spatial localisation" of the closed unity: Varela and colleagues have in mind the fact that biological systems are clearly recognisable as spatial units, distinguishable from their surroundings. Yet the criterion of being spatially localisable appears to be open to interpretation, and one could easily argue that the hydrologic cycle is a "unity recognisable in the space in which the processes exist".[5] Another relevant response would be to claim that, of course, closure is a necessary but not sufficient condition for autonomy; accordingly, those physical and chemical systems that could possibly be shown to realise closure would not be autonomous. The point is well taken but it reveals, we maintain, that we are dealing with an unsatisfactory characterisation of closure precisely because it applies to biologically irrelevant systems. As we mentioned above, and discuss in much more detail below, closure is a pivotal determination of autonomous systems, and grounds many of their distinctive properties such as, for instance, their individuation, normativity and functionality. Hence, although we would not be compelled to conclude that physical cycles are autonomous, Varela's account would indeed force us to ascribe to them many properties that we would like to apply to autonomous systems, and therefore to biological systems.

Our diagnosis concerning Varela's account of closure is that, although it points in the right direction by emphasising the fact that the organisation of autonomous systems somehow involves a mutual dependence between its components, it *fails to locate closure at the relevant level of causation.*

In our view, closure, as it is realised by autonomous systems, does not involve processes and/or reactions, as is the case for physical and chemical cycles. Instead,

[5] As a matter of fact, some authors have recently argued that the requirement for a *physical* boundary should be replaced by one for a *functional* boundary (Bourgine and Stewart 2004; Zaretzky and Letelier 2002). We agree entirely with this suggestion (see also Sect. 1.6 below), but it should be noted that functional boundaries, given that they are more general, might expose even more closure to the danger of applying to irrelevant systems. The appeal to functional boundaries should then go with a more rigorous definition of closure.

we claim that closure consists of a specific kind of mutual dependence between a set of entities having the status of *constraints* within a system.[6]

What are constraints? In contrast to physical fundamental equations, constraints are local and contingent causes, exerted by specific structures or processes, which reduce the degrees of freedom of the system on which they act (Pattee 1972). As additional causes, they simplify (or change) the description of the system, contributing to providing an adequate explanation of its behaviour, which might otherwise be under-determined or wrongly determined. In describing physical and chemical systems, two main features of the explanatory role of constraints should be emphasised. Firstly, constraints are usually introduced as external determinations (boundary conditions, parameters, restrictions on the configuration space, etc.), which means that they contribute to determining the behaviour and dynamics of a system, even though their existence does not depend on the dynamics upon which they act (Umerez 1994; Juarrero 1999; Umerez and Mossio 2013). To take a simple example, an inclined plane acts as a constraint on the dynamics of a ball resting on it, whereas the constrained dynamics do not exert a causal role in the production and existence of the plane itself. Secondly, in those cases in which some constraints are produced within the system being described, the causal relations between these constraints are usually oriented, in the sense that each constraint may possibly play a role in generating another constraint in the system, although no mutual dependence is realised.

In turn, a distinctive feature of autonomous systems is the fact that, in contrast to most physical and chemical systems, the causal relations between (at least one subset of) the constraints acting in the system generate closure. The general idea behind this account of closure is that the specificity of autonomous systems lies in their capacity for self-determination, in the form of *self-constraint*. But what does this actually mean?

Biological systems, like many other physical and chemical systems, are dissipative systems, which means, in a word, that they are traversed by a flow of energy and matter, taking the form of processes and reactions occurring out of thermodynamic equilibrium. In this respect, organisms do not differ qualitatively from other natural dissipative systems. However, what specifically characterises biological systems is the fact that the thermodynamic flow is channelled and harnessed by a set of constraints in such a way as to realise mutual dependence between these constraints. Accordingly, the organisation of the constraints can be said to achieve self-determination as self-constraint, since the conditions of existence of the constitutive constraints are, because of closure, mutually determined within the organisation itself.[7]

[6]The connection between closure and constraints has been already put forward in the work of authors like Bickhard, Christensen, Hooker, and Kauffman, mentioned in the Introduction. Similarly, substantial theoretical work has been done on this issue by various members of the IAS Research Group over the last two decades.

[7]A terminological clarification: For reasons that will become clearer later on (in particular in Chap. 5), we hold that biological self-determination (as self-constraint) implies specifically "self-maintenance" and not "self-generation". Biological systems maintain themselves but do not

As autonomous systems, biological systems do not realise some sort of "process loop" determined by a set of externally determined boundary conditions; rather, they act on the thermodynamic flow to maintain the network of constraints, which are organised as a mutually dependent network. Hence, the *organisation* that realises closure is the organisation of the constraints, and not that of the processes and reactions. What is lacking in Varela's account of closure is, we hold, the (explicit) theoretical distinction between processes and constraints, and the related ascription of closure to the organisation of constraints.

It is worth noting again that, as Varela himself has repeatedly clarified, closure (and autonomy) is by no means meant to signify the "independence" of the system vis-à-vis the external environment. On the contrary, as (Bourgine and Varela 1992) themselves explain, closure goes hand in hand with *interactive openness*, i.e. the fact that the system is structurally coupled with the environment, with which it exchanges matter, energy, and information. In our account, we ground this crucial point through the distinction between constraints and processes: while biological systems are (by hypothesis) closed at the level of constraints, they are undoubtedly open at the level of the processes, which occur in the thermodynamic flow. Autonomous systems are then, in this view, *organisationally closed* and *thermodynamically open.*[8]

Before characterising in more formal terms the fundamental distinction between processes and constraints, we shall discuss this "thermodynamic grounding" of autonomy in more detail.

1.2 The Thermodynamic Grounding of Autonomy[9]

Autonomy, as we characterise it in this book, is essentially grounded in thermodynamics. Autonomous systems, as mentioned above, are dissipative systems dealing in a constitutive way with a thermodynamic flow that traverses them.

To better understand the relevance of this statement, it is worth recalling that scientific tradition in the field of "general systems theory" has, over the last 50 years, made a clear-cut distinction between informational-organisational aspects and energy-material ones. The distinction is at the core of disciplines like Cybernetics, Artificial Intelligence, Computer and Systems Sciences, and the most recent one,

generate themselves spontaneously (as *wholes*, although of course they do generate some their functional components). In this book, we will then use "self-maintenance" to refer to the specific mode of biological self-determination.

[8] In our knowledge, (Piaget 1967) was the first author who has explicitly expressed the conceptual distinction between organisational closure and thermodynamic openness. The treatment of the distinction developed in this chapter is consistent, we think, with his own conception.

[9] Most of the ideas exposed of this section, as well as some parts of the text come from (Moreno and Ruiz Mirazo 1999).

Artificial Life, which all share the idea that considerations about the material or energy realisation[10] of a system do not affect its "organisational essence". Accordingly, although one has to include "a good deal of ancillary machinery for the real implementation of any material system" (Morán et al. 1997; see also Moreno and Ruiz Mirazo 1999), this would not be significant for modelling the organisation as such. Of course, the physical realisation of living organisation requires a certain layout of material components, interactions, and flows. Yet, the common assumption of all these approaches is that the formal and computational models can legitimately abstract from those aspects without losing their relevancy or explanatory power.

A number of abstract computational models of biological organisation have been proposed over the years, some of them at the very heart of the autonomous perspective. Just to mention a few relevant examples, Varela himself and his collaborators have developed computational expressions of autopoiesis (Varela et al. 1974; McMullin 1997; McMullin and Varela 1997), while Kauffman introduced the notion of "autocatalytic sets" (Farmer et al. 1986; Kauffman 1986). In a different computational context, Fontana's "algorithmic chemistry" generated systems (or "grammatical structures") with self-maintaining properties, expressed in the syntactical framework of the lambda-calculus (Fontana 1992; Fontana et al. 1994). All of these are models of "component production systems", sharing the basic property of self-maintenance, and their goal is to determine what is the minimal architecture of interrelations able to generate that property. In these approaches, the aspects related to energy and matter (dissipation, irreversibility, couplings, currencies, etc.) are assumed to be negligible in order to understand the principles of biological organisation.

Undoubtedly, such an abstract approach has proved to be productive for scientific research. Yet, as several authors have argued (Pattee 1977; Emmeche 1992; Moreno et al. 1994; Moreno and Ruiz Mirazo 1999), the watertight separation between "matter" (i.e. the material basis, including the energy-related aspects) and "form" (the "abstract" organisation) can be misleading when studying living systems. Rather, an adequate understanding of biological organisation should reconcile form and matter, insofar as many fundamental features of biological organisation make sense, in this view, only in relation to the conditions of their realisation in nature.

In this sense, a number of authors (Bickhard 2000; Christensen and Hooker 2000) have emphasised that what matters in this respect is precisely the fact that biological systems are also dissipative systems, so the autonomous perspective must understand biological organisation, first and foremost, in light of its "thermodynamic grounding". In general terms, this is not a new idea. Authors like (Maynard Smith 1986) and (Morowitz 1992) have already observed that the maintenance of the living

[10]The terms implementation and realisation are often used to denote a very similar meaning. However, in a strict sense, an implementation is interpreted as a kind of physical realisation of a given formal organisation which is not unique (i.e., where there are multiple possibilities of realisation of said organisation: e.g., a computer programme can be completely specified in an abstract way and then implemented in various kinds of hardware). Consequently, in this book, when we want to avoid such an interpretation, we shall use the term (physical) realisation.

state requires a constant flow of energy through the system. Either by means of an input of suitable chemical components or sunlight, energy must be supplied to the system and then (part of it) given back to the environment, typically as heat. In particular, the continuous flow of energy and matter through the system is somehow controlled by the system itself, and the way in which this control occurs is the object of disciplines such as biophysics, biochemistry, and physiology.

In accordance with these ideas, it is our contention that the autonomous perspective should integrate such dimensions into its conceptual framework. In general terms, acknowledging the thermodynamic grounding means assigning a key role to the physical magnitude *energy*, and specifically to two basic types of energy transformations (within the system and between the system and its environment): *work* and *heat*. Work is typically related to those transformations of energy that maintain or increase its "quality" (i.e. the energy gets ordered[11] – usually because it is localised[12] – as the result of the transformation and is not completely reusable a posteriori, see Atkins 1984), while heat is connected with those which "degrade" it (the energy gets dispersed, and is no longer recoverable).[13] In more technical terms, work is generated as a result of endergonic-exergonic couplings, which are not spontaneous and absorb and store energy, whereas heat is related to *exergonic* transformations, which are spontaneous and release energy.

What is to be explained, then, is how biological systems manage energy flow so as to maintain themselves, and how the very nature of their organisation is shaped to achieve this goal. If one adopts Atkins' view (Atkins 1984), then the issue can be restated as that of how biological systems succeed in constraining flows of (incoming or internally degraded) energy in order to generate work, which in turn contributes to their maintenance.[14]

The details of how living beings actually carry out this efficacious management of energy can be very complex, but, essentially, there is reasonable agreement regarding the fact that it involves the realisation of *couplings* between endergonic

[11]In a cohesive or coherent movement of constituents in the system, for instance, the classic idea of mechanical work.

[12]Typically, in a molecular bond, related to chemical work.

[13]'Heat' refers to energy that is disordered relative to the initial state of the current exothermic transition. Only in that situation does the fact of "not being recoverable any more" have a clear meaning. As we see, the concepts of work and heat are defined in terms of possibilities of energy use. But "use" here refers to a functionality which is clearly external to the system; hence, it lacks all significance without the presence of an outside observer. Insofar as living organisms are autonomous systems, we shall have to restate these concepts in such a way that they acquire meaning within the operational framework of the actual system (in Chap. 3, we shall discuss this question in detail).

[14]Actually, (Atkins 1984) does not speak about self-maintenance or biology. Rather, his book is about a general interpretation of thermodynamics, and the asymmetry between heat and work (work as a "constrained release of energy").

and exergonic processes which, in turn, requires that at least two important conditions be met (Moreno and Ruiz Mirazo 1999)[15]:

1. First, the presence of "energy intermediaries" (currencies), which enable the establishment of the couplings, so that the exergonic drive of certain reactions can be exploited or later invested in processes of an endergonic nature (typically, self-construction and repair).
2. Second, the presence of components able to modify the rates of reaction in such a way as to ensure the suitable synchronisation of the couplings. The reason is that processes, in addition to being thermodynamically feasible,[16] also follow specific kinetics. For instance, although the combustion of a glucose molecule is exergonic, its rate of reaction at physiological temperature is such that it would stay stable for ages. Accordingly, one has to take into account not only the amount of time that a reaction – or some other process – requires in order to be carried out, but also (and most especially) the time it needs in relation to other reactions with which it could become coupled. In other words, metabolism necessarily requires the *synchronisation* of a whole set of biophysicochemical processes.

Thermodynamically speaking then, biological systems are self-maintaining organisations in far-from-equilibrium conditions, which means, among other things, that their constitutive structures and relations tend to decay and cannot exist except in the presence of the continuous regeneration of the whole organisation. Self-maintenance then, occurs in spite of the continuous replacement of the material components. Indeed, biological systems may also undergo major structural and morphological changes during their lifetime, due to adaptations, accidental events (injuries, etc.) and, especially, because of development, but they keep their organisational identity whilst undergoing constant change.

1.2.1 Kauffman's Work-Constraint Cycle

(Kauffman 2000) has proposed an account that explicitly states the consequences of thermodynamic grounding on the interpretation of autonomy. His central argument consists of retrieving the classic idea of "work cycle" (as in an ideal thermal Carnot machine),[17] and applying it to the context of biochemical,

[15] On the one hand, the system must couple with some external source of energy (sunlight or chemical energy in the autotrophic case; extraneous organic matter in the heterotrophic one). On the other hand, it is also fundamental that *internal* energetic couplings take place, because this allows certain processes (of synthesis, typically) to occur at the expense of others (degradation), when in principle the former ones alone would not be spontaneously viable.

[16] Feasible in the sense that, when coupled, a *global* decrease of free energy should take place.

[17] Originally, a work cycle is a set of externally controlled processes that takes a thermodynamic system back to its initial state, giving as a result an overall production or consumption of work.

self-maintaining reactions. Specifically, he interprets these work cycles as couplings between endergonic and exergonic reactions, so characteristic in living organisms.

Based on Atkins' ideas about work, conceived, as mentioned, as a *constrained* release of energy (Atkins 1984), Kauffman argues that a mutual relationship between work and constraints must be established in a system in order to achieve autonomy in the form of a "work-constraint (W-C) cycle". The basic idea is simple yet deep: constraints are required to harness the flow of energy (in Carnot's machine, for instance, one needs the walls of the cylinder, the piston, etc.), so that the system can generate work and not merely heat (due to the dispersion of energy). In the case of systems able to determine themselves, these constraints are not pre-given, but rather produced and maintained by the system itself. Hence, the system needs to use the work generated by the constraints in order to generate those very constraints, by establishing a mutual relationship between the constraints and the work.

Accordingly, the work-constraint cycle explicitly constitutes a thermodynamically grounded self-determination, through which a system is able to self-constrain by exploiting *part* of the flow of energy and matter to generate work. Of course, the system is still thermodynamically open, and is by no means independent: it dissipates energy and matter, and has "to take in from a source" and "give away into a sink" in order to stay away from equilibrium (Morowitz 1968).

At least two implications of Kauffman's account should be emphasised.

First, the characterisation of work and constraint depends essentially on the fact that they realise the cycle. Energy is work insofar as it contributes to generating and maintaining constraints that facilitate the suitable endo-exergonic couplings. Constraints are constitutive when they are, at the same time, the *condition* (or one of the conditions) for the renewal of work in the system and the *product* of such work. As we will discuss in detail in Chap. 3, this points to the fact that, from the autonomous perspective, the meaning and value of biologically relevant determinations (such as work, constraints and the related concept of "usefulness") are grounded in the self-determining nature of biological organisation.

Second, the W-C cycle makes it clear that the autonomy of the system inherently involves the contribution of constraints. The cycle is maintained precisely thanks to the action exerted by constraints on the thermodynamic flow, which in turn regenerates the constraints. Kauffman, in our view, was one of the first authors to see not only that any understanding of biological autonomy must acknowledge its thermodynamic grounding, but also, and perhaps more crucially, that such grounding brings into focus the role of constraints exerted at the thermodynamic level.

The typical thermodynamic system would be a gas enclosed in a thermal machine (with walls that can be adiabatic or kept at constant temperature if required) undergoing successive expansions or compressions until it is brought back to its original thermodynamic state. The Carnot cycle in particular is completed through two isothermal and two adiabatic processes, producing *ideally* an amount of work that equals (or is exactly proportional to) the area limited by the lines representing those processes in a pressure-volume diagram.

Yet, although Kauffman's account is a highly relevant step towards an adequate account of autonomy, it suffers from a central weakness, namely that organisational closure implies not only the constraining action exerted on the thermodynamic flow, but also a specific *organisation* among the constitutive constraints. And the work-constraint cycle does not elaborate on the nature of this organisation.

Before explicitly addressing the characterisation of closure, let us first, in the following section, focus in precise theoretical and formal terms on the theoretical distinction between constraints and processes, as well as on two corresponding regimes of causation.

1.3 Constraints and Processes

The claim according to which biological closure is realised by the organisation of constitutive constraints acting on the thermodynamic flow requires a theoretical and formal account of the relations between the two causal regimes involved, and specifically between thermodynamic openness and organisational closure.

In this section, we provide an account of the distinction between processes and constraints (exerted on these processes). Processes refer to the whole set of physicochemical changes (including reactions) occurring in biological systems, which involve the alteration, consumption and/or production of relevant entities. Constraints, in turn, refer to entities that, while acting upon these processes, can be said to remain unaffected by them, at least under certain conditions or from a certain point of view.

We propose to ground the theoretical and formal distinction between processes and constraints in the concept of symmetry. In very general terms, symmetries refer to transformations that do not change the relevant aspects of an object: these aspects are said to be conserved under the transformations. In mathematical approaches to natural phenomena, symmetries are at the core of the constitution of the scientific objects themselves, to the extent that they ground their stability and justify the objectivity of the theories formulated to describe them. In this section, we suggest defining constraints as entities that exhibit a symmetry with respect to a process (or a set of processes) that they help stabilise. Specifically, given a process A=>>B (getting B from A), C is a constraint on A=>>B, at a time scale τ, if and only if two conditions are fulfilled. Let us discuss each of them by explaining their meanings and referring to two concrete examples, i.e. the action of the vascular system on the flow of oxygen, and that of an enzyme on a chemical reaction.

I/ The situations A=>>B and A_C=>>B_C (i.e. A=>>B under the influence of C) are not symmetrical by permutation at time scale τ.

We note $C_{A=>>B}$, those aspects of C relevant for A=>>B which, when transformed, alter A_C=>>B_C.

This condition requires that a constraint play a causal role in the target process. In formal terms, we express this by saying that the situations with and without C are not symmetrical, which simply means that they are different, even without considering the constraint itself, but just its effects on the process.[18]

Consider the vascular system. There is an asymmetry associated with the flow of oxygen when considered under the influence of the vascular system (A_C=>>B_C) or not (A=>>B), since, for instance, A_C=>>B_C occurs as a transport (canalised) to the neighbourhood of each cell, whereas A=>>B has a diffusive form. Consequently, the situation fits condition I, which means that the vascular system plays a causal role in the flow of oxygen.

Similarly, there is an asymmetry associated with a chemical reaction when considered under the influence of an enzyme (A_C=>>B_C), or not (A=>>B), since, typically, (A_C=>>B_C) occurs faster than (A=>>B).

II/ A temporal symmetry is associated with $C_{A=>>B}$ in relation to the process A_C=>>B_C, at time scale τ.

A constraint, while it changes the way in which a process behaves, is not changed by (conserved through) that same process. The second condition captures this property by stating that C or, more precisely, those aspects $C_{A=>>B}$ by virtue of which the constraint exerts the causal action, exhibit a symmetry with respect to the process ACB.[19]

Again, consider the examples. A temporal symmetry is associated with the vascular system C with respect to the transformation A_C=>>B_C, since, among other things, the spatial structure of the vascular system remains unaltered at the time scale required, for instance, to accomplish the transport of a set of molecules of oxygen from the lungs to the cells. Hence, the situation fits condition II, which means that the relevant aspects $C_{A=>>B}$ are conserved during the process.

Similarly, a temporal symmetry is associated with the configuration of an enzyme, which is preserved during the reaction.

Since they meet the two conditions, both the vascular system and enzymes can be taken as constraints within the organism.[20] All situations which fulfil conditions I and II will be expressed as C(A=>>B)τ or, in an expanded form (Fig. 1.1):

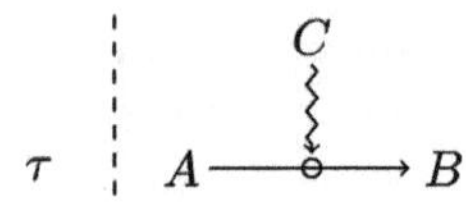

Fig. 1.1 Constraint (*Credits: Maël Montévil*)

[18]The latter precision is important because it would otherwise be trivially true that a situation AB and a situation ACB are different, because of the new object (C) that has been added. Yet, the presence of C does not necessarily change something for the objects present only in the first situation (A and B), since this depends on whether they interact with C in a relevant way.

[19]It is crucial to stress that the conservation concerns *only* these relevant aspects, while other aspects of the entity that exerts the constraint might undergo alteration, even at τ.

[20]The definition of constraint provided above is reminiscent of (and, we think, consistent with) Pattee's account of this concept (see for example, Pattee 1972, 1973). This author defines a constraint as an "alternative description" of the dynamical behaviour of a system, in which a

It is of fundamental importance to emphasise that each condition is met only *at the relevant time scales* and, in particular, that the time scale τ at which conditions I and II must be fulfilled is the same. This means that a constraint, to be such, must conserve its relevant aspects at the same time scale at which its causal action is exerted, even though it may undergo changes and alterations at shorter and/or longer time scales. Consider our two examples. The structure of the organism's vasculature does not change at those time scales at which it channels the flow of oxygen; yet, the structure of the system *does* change at greater time scales due to the effects, for example, of neovascularisation. The same holds for enzymes, which are conserved at the time scale of catalysis, while decaying and randomly disintegrating at larger scales. Moreover, enzymes are also altered at shorter time scales (since they bind with the substrate and lose or gain hydrogen, electrons or protons, etc . . .) and then restored when catalysis is achieved. In spite of the changes at longer and shorter time scales then, constraints are conserved and exhibit a symmetry at that time scale (τ) at which their causal action is exerted.[21]

In most biological cases, a constraint alters the behaviour of a system but does not lead to new behaviours. More technically, the space of possible dynamics of A_C=>>B_C is smaller or equal to the space of possible dynamics of A=>>B, each space being described at the relevant scale (the relevant scales at which the space of possible dynamics of A=>>B and A_C=>>B_C can be described may be very different from τ, and usually they are much longer, possibly infinite). In the case of the vascular system, the flow of oxygen could reach each cell at an adequate rate even in the form A=>>B, i.e. in the absence of the vascular system, from the point of view of statistical mechanics. Hence, the vascular system does not extend the space of possible dynamics of the process A=>>B. In other words, the vascular system is not required, at least in principle, for oxygen to reach the cells at an adequate rate (although the probability of the unconstrained situation occurring is extremely low, at least at biologically relevant time scales, see below). Similarly, an enzyme does not make an otherwise impossible reaction possible, but it does lead to a (possibly far) greater speed of reaction.

---

macroscopic material structure selectively limits the degrees of freedom of a local microscopic system. For an extensive discussion of Pattee's account, see also (Umerez 1994, 1995).

[21]Note that the conservation supposes that a specific time scale τ, at which the target process occurs, *is* to be specified which, in turn, requires determining when the process begins and ends. As a consequence, in those cases in which the process is continuously occurring, discretisation might be necessary to describe the constraints. Let's take the physical example of a river continuously eroding its banks. At first sight, the banks could not be taken, according to our definition, as constraints on the dynamics of the river, precisely because they are transformed by the river. But in fact this description of the system is inadequate, because it fails in specifying the relevant time scale. Although the banks are of course not conserved at the very long time scale at which the entire existence of the river can be described, their *relevant aspects* by virtue of which the river (i.e. a specifiable set of water molecules) moves from a specific point upstream to a specific point downstream in given period of time are presumably conserved during that period. Accordingly, the banks, at that time scale, fit our definition.

It is worth emphasising that the interplay between different time scales allows accounting for an apparent divergence between the idea that constraints are, in many biological cases, theoretically unnecessary, and related analyses of the role of this concept in explaining biological organisation. In particular, as (Juarrero 1999) has pointed out, constraints at work in biological systems are *enabling*, in the sense of being able to generate behaviours and outcomes that would otherwise be impossible. Now, in all those cases in which they do *not* generate new dynamics or behaviours, constraints are *limiting*: they just canalise (condition I) the constrained processes toward a specific outcome among a set of possible ones. Is there a theoretical disagreement? In fact, we think that the distinction between limiting and enabling constraints corresponds to a difference with respect to the time scale at which their causal effects are described. We maintain that, in principle, the constrained dynamics or outcomes could in most biological cases occur in an unconstrained way at the relevant (very long, or infinite) time scale; yet, at *biological* (shorter) time scales, constraints are indeed required for actually getting these specific dynamics and outcomes, because they contribute to the production of otherwise improbable (or *virtually* impossible) effects. In particular, as we will discuss in Sect. 1.5 below, each constitutive constraint of biological organisms enables the maintenance of other constrains and, because of closure, of the whole system. So, although constraints are mostly limiting at longer time scales, they can always be pertinently conceived as enabling at biological shorter time scales: in this sense, it is perfectly consistent with our account to claim that biological organisation could not exist without the causal action of constraints.[22]

Before moving on, let us discuss in some detail the theoretical and epistemological implications stemming from the distinction between constraints and processes. The central point consists in obtaining a description in which biologically relevant entities (the constraints) can be *extracted* from the thermodynamic flow to which biological systems are subject.

Condition II stipulates that the relevant aspects $C_{A=>>B}$ of the constraint are conserved, at τ, as the constrained process continues. In particular, this implies that no relevant flow of matter or (free) energy (or any conserved quantity) occurs between $C_{A=>>B}$ and A=>>B.

Consequently, we submit that constraints can be treated, at τ, as if they were *not* thermodynamic objects because, by definition, they are conserved with respect to the thermodynamic flow, on which they exert a causal action. A description of constraints in thermodynamic terms would be possible in principle, but irrelevant to

[22]At biologically relevant time scales, then, the distinction between constraints and processes roughly maps onto Rosen's distinction between *efficient* and *material* causes (Rosen 1991): constraints might indeed be said to "efficiently" produce an effect by acting, for instance, on the underlying "material" input of a reaction. In spite of this (approximate) correspondence, however, we do not adopt Rosen's terminology, which can be confusing in some respect (see also Pattee 2007 on this point), and will maintain in this book the distinction between constraints and processes. Actually, it might be argued that constraints should rather be intended as "formal" causes (see for example Emmeche et al. 2000; we also briefly discuss this question in Chap. 2).

understanding their causal role, since such a description would show that the flow between $C_{A=>>B}$ and A=>>B is at equilibrium, i.e. no alteration, consumption and/or production would be observed with respect to the constraint. A description of the causal role of constraints in terms of thermodynamic exchanges may possibly be relevant to understanding the intermediate steps leading to the effect (such as, for instance, the sequence of alterations of an enzyme during catalysis), but would be dispensable for understanding the overall effect, which does not involve a flow between the constraint and the constrained process or reaction.

Yet, according to condition I, constraints do play, at τ, a causal role in the process. How is such a role to be conceived in this framework? How can constraints be conserved and yet, at the same time, play a causal role? In our view, constraints do not produce their effects by transmitting energy and/or matter to the process or reaction, but rather by channelling and harnessing a thermodynamic flow, without being subject to that flow. Accordingly, the vasculature channels the blood flow, and the enzyme the reaction (the latter by lowering the activation energy). Even in those cases in which the constraints appear, at first sight, to transmit energy (such as, for instance, the heart which "pumps" blood), the constraint can be pertinently described as a structure which channels a source of energy (in the case of the heart, the free energy available in the cardiac cells) in order to modulate the blood flow. Again, the constraint is conserved; it exploits energy and matter to act on processes and reactions.

The central outcome of the theoretical distinction between constraints and processes consists of the claim that it corresponds to a distinction between two regimes of causation. For a given effect of a process or reaction, one can theoretically distinguish, at the relevant time scale, between two causes: the inputs or reactants (in Rosen's terms, the "material" causes) that are altered and consumed through the reaction, and the constraints (the "efficient" causes, at τ), which are conserved through that very reaction. Constraints are irreducible to the thermodynamic flow, and constitute for this reason a distinct regime of causation.

As mentioned in the previous sections, the distinction and relation between these two causal regimes is a central pillar of any adequate description of biological organisation, specifically as regards its capacity for self-determination. In the following section, we will take a preliminary step towards showing how constraints can realise self-determination in the physical domain.

1.4 From Self-Organisation to Biological Organisation

Self-determination exists in the physical and chemical domain, in the well-known form of self-*organisation*.

A classic example of self-organisation are dissipative structures (Glansdorff and Prigogine 1971; Nicolis and Prigogine 1977), in which a huge number of microscopic elements adopt a global, macroscopic ordered configuration (a "structure") in the presence of a specific flow of energy and matter in far-from-thermodynamic equilibrium conditions. In turn, the macroscopic configuration exerts a constraint on the microscopic interactions among the surrounding molecules, which contributes to the maintenance of the required flow of energy and matter, and therefore, to the maintenance of the very macroscopic configuration (Ruiz-Mirazo 2001).

A number of physical and chemical systems, such as Bénard cells, flames, hurricanes, and oscillatory chemical reactions, can be pertinently described as self-organising dissipative systems. Let us take the example of "Bénard cells", i.e. macroscopic structures that appear spontaneously in a liquid when heat is applied from below (Chandresekhar 1961). In the initial situation, in which there is no difference in temperature between the upper and lower layers, the liquid appears uniform in terms of the statistical distribution of the molecules' kinetic energy. When heat is applied, and the temperature in the lower layer is increased up to a specific threshold, the liquid's dynamics change dramatically: the random movements of the microscopic molecules spontaneously become ordered, creating a macroscopic pattern (convection cells). In each cell, billions of microscopic molecules rotate in a coherent manner along a hexagonal path, either clockwise or anticlockwise, and always in the opposite direction to that of their immediate neighbours on a horizontal plane.

Bénard cells appear when some specific boundary conditions (e.g. the heat applied from below), which exert *external* constraints on the dynamics of a given set of molecules, are imposed. Yet, once they have appeared, the maintenance of Bénard cells depends not only on these external boundary conditions, but also on the constraint exerted by the configuration itself on its surroundings. For instance, the cells *capture* surrounding water molecules in their dynamics, turning them into constituents. It is through this action that Bénard cells contribute to maintaining the flow of energy and matter traversing them.

Self-organisation, as it occurs in physics and chemistry, constitutes then a case of self-determination, described by appealing to the action of an emergent constraint on the thermodynamic flow. One needs to appeal to the constraining action of the Bénard cell itself in order to find an explanation for its own maintenance, which would otherwise be impossible on the basis of the sole properties of the boundary conditions. The macroscopic constraint determines some of the conditions required for its own existence, and then contributes to its own determination.

Is self-organisation a specific case of closure?

As we have recently emphasised (Mossio and Moreno 2010), dissipative systems realise a *minimal* form of self-determination, in the sense that they generate a *single* macroscopic structure acting as a constraint on its surrounding microscopic dynamics that, in this way, become a part of the system. Accordingly, dissipative systems make a single contribution to their own maintenance, since they contribute to maintaining the single constraint involved in the self-maintaining loop between the structure and the surroundings.

In relation to biological systems, the situation is more complex. In contrast to minimal self-organising systems, biological systems are able to exert a high number of constraints, each of them making a different contribution to the maintenance of the whole.

In doing so, they generate a network of structures, exerting mutual constraining actions on their boundary conditions, so that the whole organisation of constraints realises *collective* self-maintenance. In biological systems, constraints are not able to achieve self-maintenance individually or locally: each of them exists insofar as it contributes to maintaining the whole organisation of constraints that, in turn, maintains (at least some of) its own boundary conditions. This makes a clear-cut categorical distinction between minimal self-organisation and biological closure: while in the first case a single constraint is able to determine itself, in the second case self-determination can only be *collective*, i.e. by contributing to the maintenance of one or several other constraints, each constraint contributes indirectly to its own maintenance, because of mutual dependence (Ruiz-Mirazo and Moreno 2004).

In the following section, we will provide an explicit formal characterisation of organisational closure. Here, we would like to emphasise what is behind the distinction between self-organisation and closure in terms of the underlying complexity and interaction with the environment.

The distinction between self-organisation and closure basically involves the *takeover of (some of) the boundary conditions* required for the maintenance of the system. On the path to autonomy, closed organisations help control several environmental factors, something that requires a degree of internal complexity that simple self-organising systems do not possess.[23]

Of course, autonomous systems do depend on their coupling with the environment; as we stated earlier, autonomy is not independence. Yet, in comparison with self-organising systems, the interaction is different, the degree of dependence is lower and the system is less menaced by external changes. In cases like Bénard convection cells, for instance, small variations in the external conditions, such as the temperature gradient or the inward flow of some substrate, can provoke dramatic changes in the pattern displayed, and may even result in the complete disappearance of the structure. Generally, however, this is not what happens with autonomous systems.

Consider, for instance, the membrane. While self-organising systems are not delimited by a physical border, all biological cells possess a membrane, which not only helps to maintain an adequate internal concentration of materials and nutrients but also, because of its selective permeability, helps control their inward and outward flow. Membranes help distinguish the system from the environment, while at the same time enabling it to act on relevant factors.

The same holds in relation to those internal constraints in charge of the synchronisation of process kinetics and the establishment of global endo-exergonic couplings (see also Morán et al. 1997). The action of catalysers enables autonomous systems to take over the synchronisation of kinetics, which would otherwise depend

[23] As we will discuss at length in Chap. 5, due to the degree of complexity required by autonomous systems, these can only be historical systems (i.e. systems whose complexity has emerged through a cumulative phylogenetic process), and can by no means appear spontaneously (as dissipative structures do).

on very specific (and very unlikely) boundary conditions. Similarly, biological systems, in contrast to self-organising structures, are able to store energy so that, again, they can take over the energy supply for relatively long periods of time and be less affected by a lack of external resources.

In a word, the higher degree of complexity inherent to autonomous systems in comparison with self-organising ones corresponds to a higher degree of self-determination, because of the takeover of boundary conditions over which dissipative structures have no influence or control. The *qualitative* change from minimal (self-organisation) to collective (closure) self-determination goes hand in hand, then, with a *quantitative* increase of the underlying complexity.

One last point should be emphasised here. As we will discuss extensively in Chap. 5, the distinction between self-organisation and closure lies not just in the fact that the latter requires a higher degree of actual complexity than the former, but also in that closure allows for the *potential increase* of functional complexity. Self-organising systems, despite realising a minimal form of self-determination (we will come back to the implications of this point in Chaps. 2 and 3), are not relevant for understanding autonomy, not only because they are "too simple" and categorically different from closed systems, but also because they cannot be taken as a "starting point" for the emergence of closure and autonomy. Closure (and autonomy) is not self-organisation, and neither does it straightforwardly emerge from self-organisation.

We cannot, therefore, understand much about autonomy by looking only at self-organisation as it occurs in physics and chemistry.

1.5 Dependence

Organisational closure occurs in the specific case of mutual dependence between (at least some of) the constraints acting on a biological system. Before discussing closure as such, let us first focus on the relationship of dependence between constraints.

In the previous section, constraints were defined as entities that, among other things, are conserved (symmetrical) with respect to the thermodynamic flow. As specified above, constraints are such only at specific time scales, which means that, at other times scales, they are subject to the thermodynamic flow. In particular, at longer time scales, constraints are subject to degradation and must be replaced or repaired.[24] When the replacement or repair of a constraint depends (also) on

[24] In the case of repair the entity is maintained, while in the case of replacement it is destroyed and reconstructed. Note that the same situation can be interpreted as a case of replacement or repair following the scale at which the constraint is described: individual enzymes are replaced, while the population is repaired. This holds for all those cases (mainly at the molecular level) in which both individual and populations exert the same constraint. See the discussion about scale invariant constraint in Sect. 1.6 below.

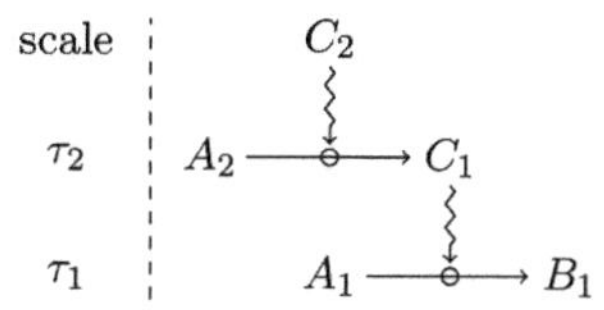

Fig. 1.2 Dependence between constraints (*Credits: Maël Montévil*)

the action of another constraint, a relationship of dependence between the two constraints is established.

As we said, a constraint C_1 is associated with a time symmetry at the scale at which it acts on the process (τ_1 below) and with respect to the relevant aspects for this process, but not necessarily at other scales (τ_2). At the same time C_1, and more precisely the relevant aspects of C_1 (as defined above), can themselves be the product of a process that, in turn, may be constrained by another constraint. This situation leads to the diagram of minimal causal dependence between constraints (Fig. 1.2).

Let us now consider a constrained process C_1 $(A_1 \Rightarrow\!> B_1)\tau_1$. Because of condition II, there is a time symmetry at scale τ_1 associated with C_1, which concerns those aspects relevant to the constrained process. At the same time, C_1 is the product of another constrained process $C_2(A_2 \Rightarrow\!> C_1)\tau_2$, at a different time scale. At τ_2, C_2 plays the role of constraint, whereas C_1 does not, being the product of the process $C_2(A_2 \Rightarrow\!> C_1)$. This situation generates dependence between constraints, where C_1 (the *dependent* constraint) depends on C_2 (the *enabling* constraint, see Sect. 1.3 above). In more general terms, we define a relationship of dependence between constraints as a situation in which, given two time scales, τ_1 and τ_2 considered jointly, we have:

1. C_1 as a constraint at scale τ_1;
2. An object C_2, which is a constraint at scale τ_2 on a process producing aspects of C_1 relevant for its role as constraint at scale τ_1 (which would not appear without this process).

As a simple example, consider the case of an enzyme acting on the reaction that it catalyses at some time scale τ. At longer scales, enzymes are subject to degradation and are replaced by the cell via the translation process, on which ribosomes and mRNA (the DNA sequence being, in turn, a constraint on mRNA) act as constraints. Hence, dependence between constraints holds between enzymes on the one side, and ribosomes and mRNA on the other.

Several important clarifications are required here.

First, the relationship of dependence that is relevant for biological closure must be a *direct* one. This specification is necessary because the definition given above would otherwise apply to a wide range of relationships between constraints, including those in which the enabling and dependent constraints are linked through very long chain of processes. Consequently, dependence would cover many biologically irrelevant situations. Hence, we restrict the relevant meaning of dependence to direct dependence, i.e. a situation in which, considering the different processes that occur at τ_2 and contribute to maintaining a relevant aspect of C_1 that depends on C_2, there is no process starting after the one constrained by C_2. For example, if we consider an enzyme formation, the maturation of the protein can be successively constrained by a chain of structures. In fact, the catalytic capacity depends directly only on the constraint acting on last process involved, which determines the conformation of the protein or, more precisely, its ability to react with other chemicals. Accordingly, the population of mRNA, as discussed above, is a constraint on the production of protein, but contributes indirectly to their conformation.

Second, dependence between constraints is logically different from dependence between processes. Indeed, at τ_2, where C_2 plays the role of constraint, the conservation of C_2 implies that no thermodynamic exchange occurs between the constraint and the constrained process, and therefore between C_2 and C_1 (see Sect. 1.3 above). In contrast, at scales other than τ_2, the relationship between constraints may involve a thermodynamic exchange, but these exchanges do not interfere with the causal dependence described at the relevant scale. At scales shorter than τ_2, exchanges are possible but irrelevant, since these exchanges would *in fine* be compensated at τ_2, at which C_2 is conserved. At scales longer than τ_2, the interaction between C_2 and $A_2 \Rightarrow\!> C_1$ might contribute to the degradation of C_2; but in the case of biological systems that degradation would be also irrelevant, since C_2 is replaced or repaired by the organisation, because of closure.

Third, in most biological cases, $A_2 \Rightarrow\!> C_1$ does not require C_2 in order to occur, at least at the very large scale of the possible evolutions of $A_2 \Rightarrow\!> C_1$. In contrast, C_2 is required, at the specific scale τ_2, to actually observe the production of C_1. The appeal to different time scales therefore allows us to circumvent the apparent contradiction between the claim that a constraint is conserved through and unaffected by the thermodynamic flow, and the fact that it depends on another constraint.

With the concept of dependence in hand, we can now turn to closure.

1.6 Closure

Closure is a specific mode of dependence between a set of constraints. In very general terms, it refers to all those cases in which, instead of having a linear chain of dependence relationships between constraints, the chain folds up and establishes *mutual* dependence.[25]

In formal terms, a set of constraints **C** realises closure if, for each constraint C_i belonging to **C**:

1. C_i depends directly on at least one other constraint of **C** (C_i is dependent);
2. There is at least one other constraint C_j belonging to **C** which depends on C_i (C_i is enabling).

[25] See also note 1 above on the conceptual relations between "closure" and "mutual dependence".

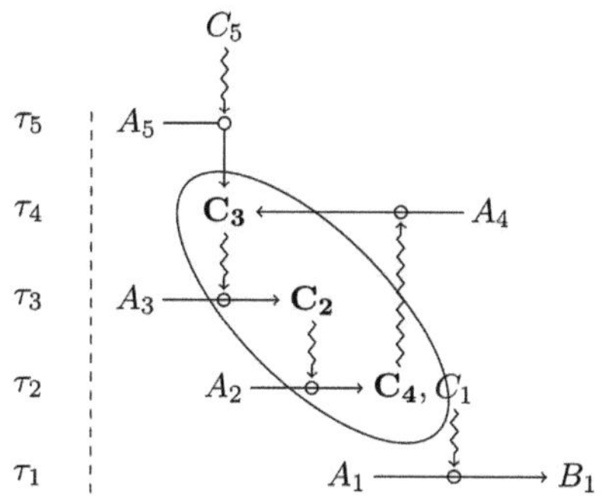

Fig. 1.3 Closure of constraints (*Credits: Maël Montévil*)

Closure refers then to an organisation in which each constraint is involved in at least two different dependence relationships in which it plays the role of enabling and dependent constraint, respectively. The network of all constraints, which fit the two requirements, is – we hold – collectively able to self-determine (or, more specifically, self-maintain; see note 5) through self-constraint.[26]

As a very general abstract illustration, consider the network of dependent constraints shown in Fig. 1.3.

In Fig. 1.3, C_1, C_2, C_3, C_4 and C_5 satisfy, by hypothesis, the definition of constraint at τ_1, τ_2, τ_3, τ_4 and τ_5 respectively. Furthermore, C_1, C_2, C_3 and C_4 play the role of dependent constraints, while C_2, C_3, C_4 and C_5 are enabling constraints. The subset that includes those constraints that are both enabling and dependent is then (C_2, C_3, C_4). The organisation constituted by C_2, C_3 and C_4 realises closure.

This definition of closure is, of course, very general and, as we will discuss in the following section, too schematic to capture the complexity of its actual realisations in biological systems. Yet, it is precise enough to illustrate some of its implications.

[26]The relations brought about by constraints responsible for closure in living systems have received two characterisations by Howard Pattee, in different stages of his work: *statistical* closure (1973) and *semantic* closure (Pattee 1982). By "statistical closure" he (1973: 94–97) means a collection of elements that may combine or interact with each other individually in many ways, but that nevertheless persists as the same collection largely because of the rates of their combination. This in turn implies a population dynamics for the elements and therefore a real-time dependence. Furthermore, the rates of specific combinations of elements must be controlled by collections of the elements of the closed set. The adjective *statistical* refers to the "selective loss of detail" of a statistical classification presents in relation to the underlying dynamics. It explains, according to Pattee, the nature and function of control constraints within a hierarchical system.

In turn, Pattee defines "semantic closure" as follows: "We can say that the molecular strings of the genes only become symbolic representations if the physical symbol tokens are, at some stage of string processing, directly recognized by translation molecules (tRNA's and synthetases) which thereupon execute specific but arbitrary functions (protein synthesis). The semantic closure arises from the necessity that the translation molecules are themselves referents of gene strings." (Pattee 1982: 333). Semantic closure is then based on the idea of symbolic records that preserve those constraints, and of how they are interpreted within the living system as a whole (Umerez 1995; Etxeberria and Moreno 2001).

Firstly, it is worth noting that, in this definition, constraints subject to closure can be interpreted both as *individual* entities or *classes* of entities. In biological systems, constraints are exerted by entities that can be described at different spatial, temporal, or organisational scales. In general, an entity exerting a constraint at a given scale also contributes to the constraints exerted, at different scales, by the larger entities of which it may be a part. For instance, an enzyme working as a catalyst in a cell could also contribute to the function of pumping blood (a different constraint) if the cell belongs to a cardiac tissue. Hence, constraints are not usually scale invariant, insofar as entities described at different scales do not exert the same constraint. Yet in some cases, a constraint might indeed be scale invariant. For instance, an individual enzyme and a group of enzymes exert the same catalytic constraint at different scales, and the same function can be ascribed both to each individual and to the group as a whole. In this case, since both the individuals and the group fit the same characterisation of "constraint" and are subject to closure, it is legitimate to claim that individual entities may be (at least to some extent) redundant,[27] since the network of causal dependencies between constraints would not be affected or altered in the event of breakdown or suppression. The constraint would still be performed by the group at a higher scale. Accordingly, the definition of closure given above covers the case in which some constraint is exerted, in a given system, by an individual entity (say: the heart) as well as those cases in which it may be exerted by an individual *or* a collection of entities.

Secondly, closure of constraints is irreducible to the underlying open regime of thermodynamic processes and changes. As discussed in Sect. 1.3, individual constraints are irreducible to the thermodynamic flow, each constraint being conserved at the relevant time scale. Hence, a reductive description of closure in terms of the causal regime of thermodynamic changes would be inadequate, since it would be unable to include constraints as such and their contribution as causal factors.[28] In particular, a description of biological organisation which does not appeal to the causal power of constraints and their closure would amount to a system constituted by a cluster of *unconnected* processes and reactions, whose coordinated occurrence would be theoretically possible at very long time scales (see discussion in Sect. 1.3), but extremely unlikely (virtually impossible) at biologically relevant time scales.[29]

[27]Scale-invariant constraints may be realised in the form of both *redundancy* or *degeneracy* of functional parts. As (Tononi et al. 1999) have pointed out, redundancy refers to the situation in which structurally similar elements produce the same effects, whereas degeneracy occurs when structurally different elements perform the same function.

[28]It is, of course, conceivable that a description of constraints might possibly be given in terms of thermodynamics, specifically as entities *that are not affected* by the thermodynamic flow. However, in this case, constraints (and hence closure) would not be reduced to a different causal regime, but simply re-described in different terms.

[29]This implication allows us to distinguish between a closure of constraints and a cycle of processes or reactions such as, for instance, the hydrologic cycle mentioned in the introduction to this chapter. In this case, the entities involved (e.g. clouds, rain, springs, rivers, seas, clouds, etc.) are connected to each other in such a way that they generate a cycle of transformations and changes between

Thirdly, as mentioned in Sect. 1.1 above, as a dimension of autonomy, closure should be carefully distinguished from independence, since a system that realises closure is a thermodynamically open one, inherently coupled to the environment. Among other things (discussed at length in Chap. 4), this implies that closure is a *context-dependent* determination, to the extent that it is always realised with respect to a set of specific boundary conditions, which include several external (and independent) constraints acting on the system. Consequently, closure does not, and cannot, include all constraints with which the system may have a causal interaction, but only the *subset* of all those that fit the definition above.

Fourthly, we understand – in accordance with Maturana and Varela – closure as a general invariant of biological organisation. Whatever its specific architecture may be, the organisation of a biological system realises closure between a subset of the constraints acting on it. Constraints subject to closure *constitute* the biological organisation and, accordingly, make an essential contribution to determining the identity of the system. Biological individuality, we think, has much to do with organisational closure, to the extent that one may conjecture that closure in fact defines biological individuality. Although this claim would require a full-fledged argument (that we leave for a future work) we do hold that, by relying on closure, the autonomous perspective clearly favours (as other authors has pointed out, see note 4) functional criteria over physical ones to define the boundaries of biological organisms. In Chaps. 4 and 6, we will make some preliminary steps to apply this view to both unicellular and multicellular organisms.

Fifthly, closure is the fundamental *principle of order* of biological phenomena, which underlies the stability of each biological system and controls the transitions and modifications that the said system undergoes over time. Many different sources of the various kinds of ordered biological patterns can of course be described; yet, what generates the distinctive order of biological organisation as a whole are – fundamentally – the principles governing the integration and coordination of its constitutive constraints in the form of closure. Accordingly, the autonomous perspective can be said to make a significant departure from molecular biology, in the sense that it advocates a shift of focus from genetic information to organisation itself as the central source of order of biological phenomena. Although, as we will discuss in more details in Chap. 5, the autonomous perspective obviously acknowledges that genetic mechanisms do play a crucial role in generating and maintaining biological organisation, it also takes an explicit *holistic* stance in claiming that the role of these mechanisms is to be understood in the light of their contribution to the whole system, the latter being governed by organisational principles.

---

them. In turn, these entities do *not* act as constraints on each other (among other reasons, precisely because they are transformed when they produce another water structure), and the system can be adequately described by appealing to a set of external boundary conditions (soil, sun, etc.) acting on a single causal regime of thermodynamic changes (see also Mossio and Moreno 2010).

Lastly, it is crucial to point out that the invariance of closure by no means implies that biological systems are not subject to variability (specifically functional variability), or that variability is not a central aspect for understanding biological systems. In our framework, closure is described by considering a temporal interval that is wide enough to encompass all constraints and their dependencies. In this sense, the organisation of mutual dependencies is described by abstracting from the physical time in which they occur. In this formal framework, the claim according to which closure constitutes an invariant of biological organisation means that a description of closure is possible for any temporal interval that is wide enough to encompass all constraints and dependencies. In other words, given a minimal interval in the thermodynamic time, closure is realised for whatever interval chosen within the system's lifetime.[30]

At the same time, biological systems may (and do) undergo changes in their organisation throughout their life. Of course, the kind of changes that are relevant here are functional changes, i.e. (as we will discuss in Chap. 3) changes involving one or more constitutive constraints. Let us emphasise that functional variations are not only a contingent fact of biological organisation but also, in many cases, a crucial requirement for adaptivity, the increase in complexity and, in the end, the long-term sustainability of life (see Longo and Montevil 2014, for an original analysis). In Chap. 5 we will discuss these issues at length. Here, it is our contention that, as biological systems undergo continuous and even inherent functional variations, *their organisation maintains closure, albeit realised in different variants*, by adding or suppressing specific constraints or sets of constraints.

Closure is a sort of organisational general invariant: it is the common property of each specific organisation that an individual system may instantiate.

1.7 A Word About Related Models

The definition of closure that we have proposed in the previous section is closely related to a number of models and proposals regarding biological organisation that have been developed during recent decades, mainly in the field of Theoretical Biology. In previous sections, we have discussed two of them in some detail, namely the theory of autopoiesis, which is also the best-known one, and the work-constraint cycle. We have also emphasised the intellectual debt that we have to Howard Pattee, specifically in relation to his work on the concept of constraint and its role in biology. In this section, we would like to say a word about some other accounts, that, of course if dealt with properly, would deserve a full-fledged analysis.

[30]See also (Montévil and Mossio 2015) for more details in this issue. In Chap. 3, Sect. 3.3.1 below we will explore the issue of the temporal boundaries of closure, when these go beyond the lifespan of an individual organism.

A first line of research has been developed by those authors who have suggested formulations of organisational closure in more chemically "realistic" terms. One example is "reflexive catalysis", defined as a gang of molecules each exerting some catalytic function so that, as a net result, the incorporation of all members of the gang is ensured by the gang itself (Szathmary 2006). A very similar concept was proposed by Stuart Kauffman in the 1980s (Kauffman 1986), under the term "catalytic closure", which refers to the mutual dependence between a set of catalysers, each of which constrains a chemical reaction that, in turn, produces at least another one of the set. Although a number of computational models and simulations of catalytic closure have been developed over the years, it should be emphasised that, to date, no chemical realisations have been obtained, which shows to what extent even minimal instantiations of closure require a non-trivial degree of complexity.

In this sense, a very relevant contribution has been made by the Hungarian biologist Tibor Gánti through his model of the *chemoton* (Gánti 1975/2003). The chemoton consists of three functionally dependent autocatalytic subsystems: the metabolic chemical network, the template polymerisation and the membrane subsystem enclosing them all. The correct functioning of the chemoton relies on the precise stoichiometric coupling of the three subunits. The most important of these cycles is the first one because it transforms the chemical energy of nutrients into useful work and constitutes the material support for the other two subsystems. The compartment isolates the autocatalytic subsystem, ensuring an adequate concentration of components and making a certain selection in the transport of matter between the environment and the system. As in the case of autopoiesis, the chemoton creates its own membrane: the metabolic cycle generates not only more intermediates of the cycle but also components of the membrane. Moreover, the chemoton includes the capacity for self-reproduction, since the dynamics of both the compartment and the inner components evolve, doubling their initial value and leading to the subsequent division into two identical chemotons. Lastly, Gánti added a third subsystem – the "template cycle" – to ensure a kind of "control" or "regulation" of the other two dynamically coupled subsystems. It is the length of the polymer that matters in the regulation of the other two subsystems, since it affects the replication rate. The role of the template has nothing to do with any informational control of present-day cells; rather, it is more like a kind of "buffering" system, acting as a "sink" soaking up the waste products of the metabolism, and so affecting the metabolic rate (we shall discuss this point further in the next section). In sum, the combination of the three subsystems gives rise to what Gánti characterises as a supersystem, displaying biological features. In this way, Gánti defines a threshold of minimal tasks and avoids trivial forms of self-maintenance (Fig. 1.4).

Although Gánti does not explicitly refer to constraints closure, the chemoton is undoubtedly an example of closure of constraints because (at least) the membrane that contains the reaction network and the catalysts driving these reactions operate as constraints, and they depend on each other. Moreover, since the length of the template also acts by affecting the rates of other basic processes, it fits our characterisation of a constraint as well. And in turn, this constraint is also dependent on the other constraints. Hence, the chemoton fulfils the criteria of an

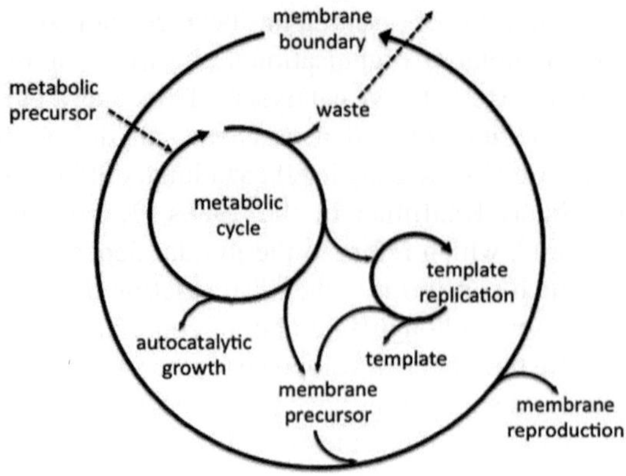

Fig. 1.4 Scheme of Gánti's Chemoton with the three coupled cycles *(credits: Juli Peretó)*

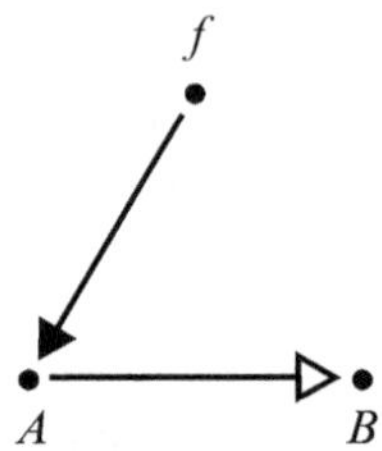

Fig. 1.5 Rosen's distinction between efficient and material causes

organisationally closed system, and provides very relevant insights into the degree of chemical complexity that even its minimal realisations must attain in order to show relevant biological features.

The second line of research is that established by Robert Rosen, and currently being developed by several authors (see for instance Letelier et al. 2003, 2006; Cárdenas et al. 2010; Piedrafita et al. 2010). Rosen's account is complex and profound, and aims to provide a conceptual, theoretical, and formal characterisation of the general principles of biological organisation (as well as of the modelling relationship itself, although we will not discuss this aspect of his work here). Although he was probably not the first author to have used the term "closure" to refer to a distinctive property of biological systems, he was certainly the first one to have explicitly seen and claimed that a sound understanding of closure in biological organisation should make the distinction between two causal regimes at work in biological systems and should locate closure within the relevant regime. In this sense, we acknowledge the intellectual debt we owe to Rosen's work, and see our work, in many ways, as an attempt to further develop his ideas and insights.

As Rosen's account has been developed over 40 years, we refer here only to his latest contributions, and in particular to his book: *Life Itself* (Rosen 1991). As described in that volume, his account is based on a rehabilitation and reinterpretation of the Aristotelian categories of causality and, in particular, on the distinction between efficient and material cause. Let us consider an abstract mapping f between the sets A and B, so that f: A=>>B. Represented in a relational diagram, we have Fig. 1.5.

Fig. 1.6 Rosen's closure to efficient causation

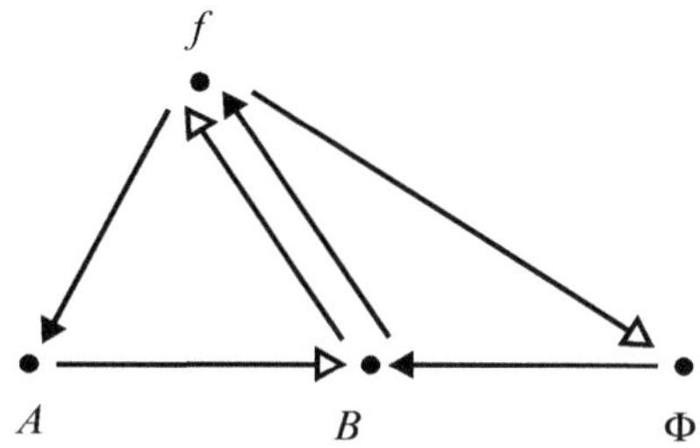

When applied to model natural systems, Rosen claims that the hollow-headed arrow represents material causation, a flow from *A* to *B*, whereas the solid-headed arrow represents efficient causation exerted by *f* on this flow.

Rosen's central thesis is that "a material system is an organism [a living system] if, and only if, it is closed to efficient causation" (Rosen 1991: 244). In turn, a natural system is closed to efficient causation if, and only if, its relational diagram has a closed path that contains all the solid-headed arrows. According to Rosen, the central feature of a biological system is the fact that all components having the status of efficient causes are materially produced by and within the system itself. At the most general level, closure is realised in biological systems between three *classes* of efficient causes corresponding to three broad classes of biological functions, which Rosen denotes as *metabolism* (f: A=>>B), *repair* (Φ: B=>>f) and *replication* (B: f=>>Φ) (Fig. 1.6).

By providing a clear-cut theoretical and formal distinction between material and efficient causation, Rosen, as mentioned above, explicitly distinguishes between two coexisting causal regimes: closure to efficient causation, which grounds its unity and distinctiveness, and openness to material causation, which allows material, energy, and informational interactions with the environment. Clearly, the distinction between constraints and processes maps onto the distinction between efficient and material causes. As a matter of fact, as Pattee discusses in a recent paper (Pattee 2007), in his previous work Rosen himself used to use a terminology that was closer to the one adopted in this book.

An analysis of Rosen's account in all its richness would far exceed the scope and limits of this book.[31] Here, we would simply like to mention a specific point about which we believe Rosen has proved particularly insightful. As mentioned earlier, closure to efficient causation is realised by and between very general classes of functions, not just between individual constraints. Accordingly, Rosen's closure occurs at a *higher level of description* with respect to ours, and the relevant question is why Rosen chose to define closure at that level, and what are the implications of that choice. To the best of our knowledge, no clear answers have

[31] We made a contribution in Mossio et al. (2009a), in which we analysed one of Rosen's claims, according to which closure to efficient causation has non-computable models. (Cárdenas et al. 2010) offers a detailed reply to our analysis.

yet been provided to these questions, although in recent times, some studies have taken important steps in this direction. In particular, Letelier and co-authors have published an analysis of Rosen's account in which they propose an interpretation of its central features, and focus in particular on the biological meaning of the classes of function subject to closure (Letelier et al. 2006). In their view, Rosen's labels are somehow misleading, and they suggest using "metabolism", "replacement", and "organisational invariance". They discuss at length the last class of functions, and claim that Rosen's central result was the mathematical demonstration that a system endowed with metabolism and replacement functions can also be inherently organisationally invariant.

Without entering into any mathematical detail, we would simply like to emphasise that, in our view, Rosen's account touches on a crucial issue related to the principles of biological organisation, namely the fact that we should be able to understand its invariance through time, and that said understanding requires an appeal to higher-order closure and organisation. The connection between the invariance (or at least stability) of organisation and the hierarchical nature of constraints and closure is, we believe, a key topic for future research in the field of theoretical biology, especially from the autonomous perspective. In the following section, we make some preliminary headway in this direction.

1.8 Regulation

The characterisation of closure offered in the previous section is extremely general, and aimed at covering all its possible concrete realisations in nature. Any system realising closure, we submit, has to fulfil the above characterisation. Yet concrete realisations of closure require a minimal degree of complexity in order to be not only possible, but also biologically relevant. Indeed, not all closed systems belong to the biological domain, since some viable closed networks may not possess biologically relevant properties or features.

In this section, we focus specifically on this issue, and try to clarify how such "biologically relevant" properties and features should be understood. Furthermore, it will be our contention that those realisations, which are complex enough to be biologically relevant, provide the theoretical groundwork for understanding *metabolism*, interpreted within the framework of the autonomous perspective.

Under what conditions, then, can actual instantiations of closure be taken as "metabolic"? Usually, the simplest realisations of closure are conceived in the chemical domain, in the form of catalytic closure, as briefly discussed in the previous section. The simplest option, of course, is to claim that any minimal chemical network is *ipso facto* metabolic, providing it realises catalytic closure. As a matter of fact, a number of physicists and chemists, typically interested in the origins of life, have characterised minimal metabolism in a very simplified way,[32] as a closed network of reactions, typically driven by pre-enzymatic catalysts.[33] Interestingly, these networks are usually referred to as "proto-metabolisms[34]" meaning that, although they are very simple, there is a fundamental organisational continuity between them and fully-fledged metabolisms. What constitutes metabolisms is already there, although *in nuce*, in proto-metabolisms.

Now, present-day metabolisms, however simple, show a rich organisational diversity. Actually, in the prokaryotic world, the diversity of metabolisms is truly astonishing. Therefore, it is only logical to consider that the concept of metabolism should somehow imply the capacity to harbour, at least potentially, an indefinite organisational diversity. In fact, the kind of organisation that constitutes the core of biological systems implies a capacity to potentially enlarge indefinitely the number of constraints and therefore, its complexity; otherwise the system would not have the capacity to evolve in an open way.

Yet, as we shall explain, minimal conceivable realisation of organisational closure, as we have already described it, does not ensure this capacity. The central point is that systems realising closure do not necessarily possess the capacity to compensate for variations, be they internally or externally generated. Consequently, variations (such as, for instance, changes in component concentrations in one reaction) can affect the output of a specific constitutive constraint, which in turn may affect the structure and activity of other constraints, and so on. Because of this "transmission of variation", due to the closure between constraints, the organisation may progressively "drift" and, most likely, become disrupted after a short time. Moreover, given the "delicate balance" (see below) between the constituents, the more the organisational complexity increases, the more crucial the capacity to compensate for variations becomes. As an example, take the case of Gánti's chemoton, which can be disrupted even by slight variations due to perturbations in the environment. As (Bechtel 2007) pointed out:

---

[32]It should be mentioned that there is another conception of minimal metabolism, typically put forward by biochemists (see for instance Gil et al. 2004), according to which it is the characterisation of "minimal genomes" through the simplification of existing ones, under the assumption that their associated metabolic networks will drastically reduce the complexity of extant metabolisms. Here, minimal metabolisms are still "genetically-instructed metabolisms", similar (although highly simplified) to those realised by fully-fledged living organisms. As discussed by (Morowitz 1992) and (Morange 2003), one of the problems of this conception seems to be that, since metabolic simplicity depends on the environment, it is highly problematic to elaborate shared criteria to determine what *the* minimal metabolic network actually is.

[33]For example, (Eschenmosser 2007: 311), writes that "another type of reaction loop that can emerge as a consequence of the exploration of a chemical environment's structure and reactivity space is one that, driven by the free energy of starting materials, connects intermediate products (substrates as opposed to catalysts) in a cyclic pathway: such a cycle is referred to as autocatalytic *metabolic cycle*."

[34](De Duve 2007) uses the term "proto-metabolism" to denote those chemical networks driven by catalysts that, whatever their nature, cannot have displayed the exquisite specificity of present-day enzymes and must necessarily have produced some sort of "dirty gemisch".

> Imagine the environment changed so that a new metabolite entered the system which would react with existing metabolites, either breaking down structure or building a new additional structure. This would disrupt the delicate balance between metabolism and membrane generation that Gánti relies on to enable chemotons to reproduce. What this points to is the desirability of independent control of different operations within the system (p. 299).

As we will see in Chap. 5, variations and the transmission of variations play a fundamental role in enabling the generation of novelty in biological systems, and contribute to their long-term evolution. Yet, as we claimed in Sect. 1.6 above, variation cannot play any evolutionary role unless it can be governed by biological organisation, in order to guarantee its stability while at the same time enabling it to integrate novelty. Biological organisation must be able to handle variations, and then conserve closure, otherwise it would be extremely fragile and its realisations in the natural world would hardly move beyond a very low level of organisational complexity. Any perturbation would be more likely to drive the system to disruption than to result in an increase of complexity. What is then required for biological organisation not only to remain stable in the face of perturbations, but also be able to increase its complexity? The answer is, we submit, *regulation*. Biological autonomy requires regulated closure.

When a set of constraints realise closure, the collective maintenance of the organisation lasts for as long as the activity of each constraint stays within admissible ranges and, moreover, adequate external boundary conditions (on which, as we will see in Chap. 4, the system has a partial influence) persist. If a variation occurs,[35] the system drifts and (possibly, see below) collapses unless it possesses some additional capacity enabling it to respond to the variation, and compensate for its effects.[36] How can closed organisations handle deleterious variations? Let us leave aside those local cases that we could label "local robustness", i.e. the capacity of a *single* constraint or structure to compensate for a perturbation, without altering its behaviour or its causal effects. In this case, the perturbation is handled and compensated for locally, and does not produce a variation affecting the relations which exist between the constitutive constraints; in this sense, it is irrelevant from the point of view of the whole organisation, and as such, is not included in our discussion. In contrast, we focus on those variations that do alter the activity of local

[35] We focus here on deleterious variations, i.e. variations that do not lead to new viable organisations and would disrupt the system if not compensated.

[36] It should be noted that, in some cases, variations may be neutral with regard to the self-maintenance of the system: in spite of the variation, the system may drift, but closure is conserved. And it might be the case that the biological system exerts a form of compensation even on this kind of harmless variation, counteracting its effects. In what follows, however, we shall not discuss these forms of compensation because they are negligible with respect to maintaining closure, which is, after all, the main reason for requiring regulation. Regulation will then specifically be characterised in relation to cases of "deleterious" variations that disrupt closure: a gap is generated between the conditions of existence and the activity of the system, which is no longer able to meet those very conditions of existence, and is therefore destined to collapse.

constraints, thus calling for a response by the global organisation (Barandiaran et al. 2009). Regulatory capacities are about these global responses.

In order to understand how these systems deal with perturbations, we shall introduce a distinction between two general ways self-maintaining systems deal with environmental variations and/or, in general, perturbations. Examples of the first way can be grouped under the general label *constitutive stability* against some range of perturbations. The second way is *regulation (or adaptive regulation).*

1.8.1 Constitutive Stability

Constitutive stability is the capacity of the whole biological organisation to respond to, and compensate for, variations, thanks to the specific structure of the network of constraints, which might for instance instantiate loops of negative feedback. A variation affecting a given constraint can propagate within the system, and produce the variation of one or several other constraints that in turn compensate for the initial one. As a result, the system is stable, homeostatic.

One way of thinking about a primitive form of constitutive stability was proposed by (Deamer 2009), in a discussion concerning the origin of Life:

> No one has yet attempted to develop an experimental system that incorporates all of the above components and controls, so one can only speculate about how control systems might have developed in early forms of life. One obvious point in the network offers a place to start. Small nutrient molecules must get across the membrane boundary, and so the rate at which this happens will clearly control the overall process of growth. I propose that the first control system in the origin of life involved an interaction of internal macromolecules with the membrane boundary. The interaction represents the signal of the feedback loop, and the effector is the mechanism that governs the permeability of the bilayer to small molecules. As internal macromolecules were synthesized during growth, the internal concentration of small monomeric molecules would be used up and growth would slow. However, if the macromolecules disturbed the bilayer in such a way that permeability was increased, this would allow more small molecules to enter and support further growth, representing a positive feedback loop. The opposing negative feedback would occur if the disturbed bilayer could add amphiphilic molecules more rapidly, thereby reducing the rate of inward transport by stabilizing the membrane. This primitive regulatory mechanism is hypothetical, of course; however, it could be a starting point for research on how control systems were established in the first forms of life.

Another example of constitutive stability is the chemoton itself. In this kind of system, disequilibrations are compensated for through the mutual interaction of stoichiometrically-coupled subsystems, that act by affecting the concentrations of the products of the distinct cycles, in such a way that the rates and speed of reactions inside the system are collectively determined. Moreover, the capability of the template subsystem to activate the production of the membrane once a certain threshold of concentration in the product of the metabolic subsystem (determined by the length of the template polymer) is reached, constitutes a mechanism of delay or damping of internal disequilibrations. This mechanism is analogous to a water reservoir that establishes a threshold of concentration of the side product

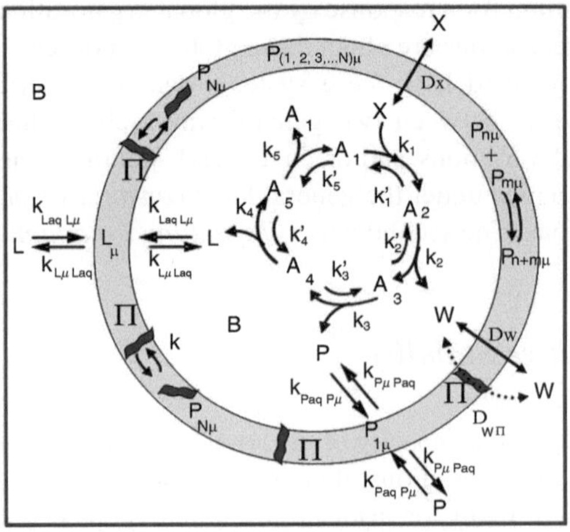

Fig. 1.7 Schematic graph of a minimal self-maintaining compartmentalised organization based on the complementarity between an internal autocatalytic reaction cycle and the self-assembly processes that make up the membrane (from its lipidic and peptidic building blocks). Peptides inserted in the membrane ensure the constitutive stability of the system by opening channels when internal pressure attains a critical threshold (Source: Adapted with permission from (Ruiz-Mirazo and Mavelli 2007). Copyright 2007 Springer-Verlag)

of the metabolism (used in the replication of the template), for the activation of the production of the membrane. This threshold is dependent on the length of the template polymer. Its role would consist of the system responding to an increase in internal pressure by building more membrane and thus avoiding an osmotic burst.

A final and interesting example is the recent model proposed by (Ruiz-Mirazo and Mavelli 2007, 2008). This is a self-reproducing vesicle whose membrane consists of both fatty acids and small peptides, such that the "mechanical" dynamics of the membrane are operationally coupled to the chemical dynamics of the internal autocatalytic network. The system realises control operations so as to maintain a steady state: when the osmotic pressure reaches a certain threshold, peptides in the membrane open channels; and this happens because, due to the elastic tension (a mechanical process), peptides inserted in the membrane adopt the conformation required to become waste-transport channels, thus enabling a faster release of the waste molecules and, consequently, a decrease in osmotic pressure differences (Fig. 1.7).

Constitutive stability requires not only that a set of constraints be mutually dependent but also that their activity can be modulated by specific (internal or external) perturbations in a way that preserves closure. What matters for our discussion here is that the response to the perturbation consists of a chain of changes affecting the constitutive organisation that, because of the architecture of the network of constraints, compensates for the initial variation. This means, in particular, that constitutive stability *does not require that we appeal to a different subset of constraints* specifically in charge of handling deleterious variations; rather, the variations of the constitutive organisation itself, induced by the perturbation, are sufficient.

Constitutive stability is *conservative*, and brings the system back to the same organisation that was in place when the perturbation occurred. The only way in which the system can change is by moving to a different organisation (i.e. a different "regime of self-maintenance") through the establishment of new stable dependencies between constraints, but this would be just a transition between different regimes, determined by the perturbation, and not an increase in overall complexity. Accordingly, although it enables the system to handle certain deleterious variations, constitutive stability is not a relevant starting point for the increase of organisational complexity since it does not enable the system to explore different regimes of closure.

1.8.2 Regulation[37]

The second way of handling perturbations is regulation, which is based on a qualitatively different form of organisation. Regulation requires that the closed organisation possesses a set of constraints exclusively operating when closure is being disrupted by a deleterious variation. The role of these constraints consists of re-establishing closure and bridging the gap between the activity of the system and its conditions of existence, by modulating (and possibly modifying) the constitutive organisation itself and/or its interaction with the environment. By definition, therefore, regulatory constraints are different (and complementary) with respect to constitutive ones: they do *not* contribute to the maintenance of closure in stable conditions (while constitutive ones do) but, when closure is being disrupted, they govern the transition towards its re-establishment (while constitutive ones do not).

As an example, consider the lac operon system, which regulates the metabolism of lactose in bacterium *E. coli*. In normal circumstances, *E. coli* metabolises the glucose taken in the environment. When the level of glucose becomes very low, and lactose is abundant, a mechanism called lac-operon is activated: the detection of lactose disinhibits the expression of a cluster of genes that enable lactose metabolism. In circumstances in which the availability of glucose is constant (that we take here as a set of constraints acting on a sequence of changes of metabolic pathways) these genes do not contribute to the maintenance of the organisation. The cluster of genes remains dormant and would not be included in the characterisation of the current closed organisation of constraints: in those conditions, nothing else in the system depends on the lac operon for its own maintenance. In turn, the lac operon becomes operational when a perturbation (the decrease of glucose levels) occurs and the maintenance of the organisation is menaced: the lac operon re-establishes closure by modifying the constitutive organisation (which shifts from glucose to

[37]The content of this section owes a lot to preliminary discussions with Leonardo Bich and Kepa Ruiz-Mirazo. See Bich et al. (forthcoming) for details.

lactose metabolism), and bridges then the gap between the activity of the system and its conditions of existence. Accordingly, the lac operon mechanism is regulatory.

A major implication is that *regulatory constraints are not subject to constitutive closure*, precisely because in a stable situation in which no deleterious variations occur they are not enabling (see Sect. 1.5. above), i.e. there is no constraint that depends on them. In turn, we claim that regulatory constraints are *second-order* constraints that, unlike constitutive ones, exert their causal actions *on changes of other constitutive constraints* of the organisation. In the case of the lac operon, for instance, the regulatory mechanism governs the transition from glucose to lactose metabolic pathways, which themselves consists of a set of constraints acting on underlying chemical reactions. In particular, in accordance with the definition given in Sect. 1.3 above, regulatory constraints exert a causal action, at time scale τ, on a change related to one (or a set of) other constraint(s), while being conserved through such change at τ. As with any constraint, their causal role at τ is intimately linked to their conservation, since the properties that are conserved are precisely those that provide them with specific causal powers. The lac operon mechanism is conserved at the time scale at which the shift from glucose to lactose metabolic pathways occurs.

The fact that regulatory constraints are not subject to first order closure is then what allows distinguishing them from the first order organisation, i.e. for distinguishing the "regulating system" from the "regulated system" in a principled way. One important implication is that, since they do not participate to constitutive closure, regulatory actions are *triggered* when the relevant (class of) perturbations occur. Therefore, a conceptual distinction can be made between the "constitutive" processes that maintain the regulatory functions (as for instance those which maintain the cluster of dormant genes responsible for the glucose/lactose switch in the lac operon case) and the processes (or changes) that trigger their action (as the increase of lactose and decrease of glucose in the environment). The triggering processes, ultimately due to an external or internal perturbation, may take many specific forms: in particular, it is worth noting that they are in many cases completely *distinct* from the constitutive ones, as for the lac operon. As a consequence, regulatory constraints realise a sort of *decoupling* from the constitutive organisation not only with respect to their effects, but also with respect to their dependence from the constitutive organisation for their triggering. Not only does regulation not contribute to constitutive closure but typically it is not even triggered by (changes of) processes involved in the constitutive closure. Such a decoupling of regulatory constraints vis-à-vis first-order organisation is, we will point out below, what allows them to play a crucial role in the increase of complexity (Fig. 1.8).

The regulatory subsystem (R), when activated (R/P) by a triggering perturbation (P), governs the transition from one constitutive organisation ($C_1 \ldots C_n$) to another one ($C_1' \ldots C_j$). In this specific case, the difference between the two constitutive organisations consists in the replacement of the constraint C_n with C_j.

At this point, a possible objection could be the following: if regulatory constraints are not subject to closure, can we still claim that they are part of the system? Does this characterisation imply that they are *external* to the organisation? We reply by claiming that, if one adopts a broader view, regulatory constraints can

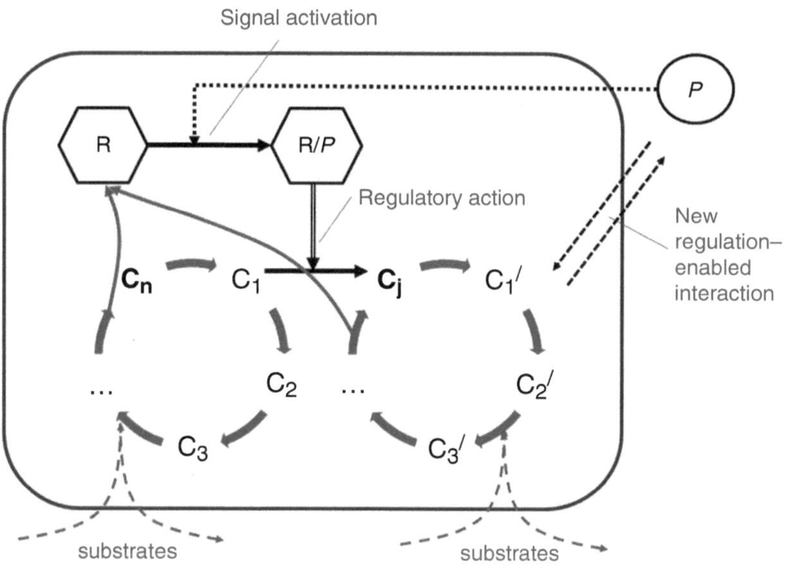

Fig. 1.8 Regulation (*Credits: Leonardo Bich*)

be shown to be both dependent and enabling at the same time, and therefore still subject to closure. In particular, they govern the *transition* between two organisations, the one whose closure is collapsing (the glucose-based one, in the example of the lac operon), and the one that they contribute to establishing (the lactose-based one): regulatory constraints depend on the (constitutive constraints of the) former, and enable the (constitutive constraints of the) latter. We argue that, accordingly, regulatory constraints are subject to a *second-order closure* between both themselves and the whole *set* of organisations among which they govern the transitions. The closure is of second-order because it is realised by a second-order organisation constituted by its set of regulatory constraints, on the one hand, and the set of available instantiations (one of which is enabled/channelled at a given moment) on the other. In other words, this second-order organisation consists of the set of available constitutive regimes of a closed organisation *given* a specific set of regulatory constraints and a set of deleterious variations to which the regulatory constraints are specifically sensitive.

Usually, these changes of regime are reversible and the second-order organisation may instantiate a previously collapsed first-order organisation if a new variation (or an end to the previous variation) were to activate regulatory capacities in this direction (in the case of the lac-operon, this would occur if the availability of lactose decreased, and that of glucose increased again). While the first-order organisation of constraints allows a modulation of the basic physicochemical processes, regulatory second-order constraints modulate the structure of the (first order) organisational closure. To take another example, a typical mechanism of regulation involves allosteric enzymes, which have two (or more) binding sites and the capacity to switch between different metabolic paths. Accordingly, regulation responds to variations by inducing the (reversible) switch from one first-order instantiation to another: the system changes its organisation to maintain closure, which makes it not only robust, but also *adaptive.*[38]

In Chap. 3, we will argue at length that closure provides a naturalised ground for functionality, and its normative and teleological dimensions. Constraints subject to closure – we will claim – correspond to biological functions, and the conditions of existence of the closed organisation are the norms that they are supposed to satisfy. Here, let us emphasise that the idea according to which regulation is subject to second-order closure implies that it is subject to *second-order norms*, i.e. the norms generated by the conditions of existence of the second-order organisation. Accordingly, regulatory constraints are supposed to contribute to its maintenance by inducing the realisation of one of its possible instantiations (in the example of the lac operon, the possible instantiations being the glucose-based and the lactose-based organisations), according to the specific perturbation that affects the system. To the extent that the effects of regulation involve a shift from one specific constitutive organisation to another, and then from one closure to another, it follows that *regulation modifies first-order norms according to second-order norms.* More subtly: not only can regulatory constraints modulate the inherent norms of the organisation (as constitutive constraints do when the organisation varies for some reason) but, crucially, that modulation is *itself* teleological and normative: regulation is then, in the autonomous perspective, *functional modulation.* And metabolisms are those organisations realising regulated closure.

1.8.3 Regulation and the Increase of Complexity

At this point, we can come back to the initial question: why is regulation, characterised in this specific way, a relevant starting point for explaining not only stability, but also the increase of biological complexity? Or, as we framed the issue, why should we take regulated closure, and not just constitutive stability, as a characterisation of metabolism?

The central point is that regulation allows stability while enabling the increase of complexity, because second-order constraints are decoupled from the constitutive organisation, and therefore less affected by the perturbations impinging on it.

[38] As Di Paolo puts it, adaptivity is "a system's capacity to regulate, according to the circumstances, its states and its relation to the environment with the result that, if the states are sufficiently close to the boundary of viability, (1) tendencies are distinguished and acted upon depending on whether the states will approach or recede from the boundary and, as a consequence, (2) tendencies of the first kind are moved closer to or transformed into tendencies of the second and so future states are prevented from reaching the boundary with an outward velocity" (Di Paolo 2005: 438). For a detailed discussion, see also (Barandiaran et al. 2009); (Barandiaran and Egbert 2013).

What does this imply? In the case of constitutive stability, as (Christensen 2007) has also argued, achieving compensation depends on propagating changes through many local interactions within the organisation: this means that the time taken to achieve it can be long and, crucially, increasingly long as the size of the system increases. Regulation instead allows a decoupled subsystem to induce the appropriate collective pattern in a more rapid and efficient way. The modulation of the system is more efficient if, instead of modifying the very constitutive organisation, it can control the switches between available regimes, through a dedicated mechanism able to cope with specific perturbations. "Freed" from first-order organisation, at least at relatively short time scales, higher-order constraints can be left to "spontaneous" dynamics and allowed to explore higher degrees of organisational complexity, providing that, at longer time scales, it contributes to maintaining second-order closure, as explained above.

Let us illustrate how this can happen. Imagine a modified chemoton in which some modular components acquire functional tasks due to the specific sequence of their constitutive modules, as (Griesemer and Szathmáry 2009) have argued. If, instead of just one type of molecule being combined into the sequence of this modular component (say, a short polymer), two or more types constitute the building blocks, then the system will exhibit both a composition of its building blocks in specific concentrations and a sequence. While the concentrations, like other features of the chemoton, will depend on specific stoichiometric relations, the *sequence* is, stoichiometrically speaking, a "free" property. In turn, this stoichiometrically decoupled property can possibly be linked to component operations in the chemoton, so as to control them. In this situation the sequence, which does not participate directly in maintenance, takes over regulatory functions.[39] This allows a free exploration of the reaction space and, furthermore, once a regulatory hierarchical order has been shown to be possible, this in turn opens up the path to the creation of a new higher regulatory orders, and so on, indefinitely. As (Mattick 2004) has pointed out, what really enables the increase in complexity in biological evolution is not so much the capacity to generate a rich variety of elements, but rather the capacity for functional and selective control. The preliminary account of regulation from the autonomous perspective developed in these pages points to the same direction.

[39](Bechtel 2007) has pointed out the same argument, which he states as follows: "if control is to involve more than strict linkage between components, what is required is a property in the system that varies independently of the basic operations. The manipulation of this property by one component can then be coordinated with a response to it by another component so that one component can exert control over the operation of the other component " (Bechtel 2007: 290).

1.8.4 Towards Autonomy

Before concluding this chapter, an important clarification concerning the conceptual connection between regulation and autonomy is needed. In our view, regulation represents a qualitative transition on the path towards autonomy, not only because it enables an increase in functional and structural complexity, but also because the system has internalised constraints which are able to modify norms (those belonging to first-order organisation) in order to preserve its own existence. When closure is regulated, the system not only generates intrinsic norms but, as we emphasised, modulates these norms in order to promote its own maintenance; and this does not happen randomly, but in accordance with second-order norms. This means that self-determination has a stronger sense here: it is not just the generation of intrinsic norms, but also their submission, in accordance to (other) norms, to the maintenance of the system. And only the realisation of a *hierarchy*[40] of (at least) two orders of closure allows this distinction.

Hierarchical (regulated) closure, however, is not autonomy. As we will discuss in the following chapters, and particularly in Chap. 4, autonomy also implies the integration of an interactive dimension, which deals with the relations existing between the biological system and its environment. And we will see that, since regulation may concern both internally and externally-generated perturbations, the system can also exert a causal influence on its environment in order to promote its own maintenance: this is what we call adaptive agency.

[40]It is worth emphasising that, from the autonomous perspective, biological organisations might be hierarchical with respect to both *orders* and *levels* of closure. Chapter. 6 addresses explicitly this conceptual distinction.

2
Biological Emergence and Inter-level Causation

Whether adequate explanations in biology require appealing to an emergent and distinctive causal regime seems to have an obvious positive answer, insofar as biological systems evolve by natural selection (Mayr 2004: 31). Yet, as Wesley Salmon has pointed out (Salmon 1998: 324), one may distinguish between etiological explanations, which tell the story leading up to the occurrence of a phenomenon, and constitutive explanations, which provide a causal analysis of the phenomenon itself. Accordingly, whereas this goes without saying for etiological explanations, there seems to be no obvious answer to the question of whether a constitutive explanation of biological systems would also appeal to a distinctive regime of causation, emergent from and irreducible to that at work in natural physical and chemical systems.

What does the autonomous perspective have to say with respect to this issue? The view according to which closure constitutes a distinctive feature of biological organisation seems to require the adoption of a non-reductivist stance, according to which biological systems realise a regime of causation that is irreducible to (and then distinct from) those at work in other classes of natural (i.e. physical and chemical) systems. Indeed, in Chap. 1, we explicitly claimed that constraints and closure are irreducible to the thermodynamic flow on which the causal action is exerted, because of their conservation at the relevant time scales. Accordingly, the autonomous perspective seems to clearly advocate an emergentist stance with respect to constitutive explanations in biology. Yet, this claim calls for an adequate philosophical justification: which properties do make closure of constraint irreducible and emergent? What is the precise account of emergence invoked here? In this chapter, we aim to develop these issues in some details, by situating the autonomous perspective within the context of the more general discussion on emergence and reduction.

The ideas presented in this chapter, as well as most parts of the text, were originally presented in (Mossio et al. 2013).

A. Moreno, M. Mossio, *Biological Autonomy*, History, Philosophy and Theory of the Life Sciences 12, DOI 10.1007/978-94-017-9837-2_2

Similarly, it is currently unclear whether or not closure involves inter-level causation. At first sight, it seems obvious that closure inherently relies on the causal interplay between entities at different levels of description: the integrated activity of lower-level constituents contributes to generate the higher-level organisation, and the higher-level organisation plays a crucial role in maintaining and regenerating its own constituents, as well as controlling and regulating their behaviour and interactions. As a matter of fact, Moreno and Umerez did argue in a previous contribution that inter-level causation is a fundamental aspect of biological organisation (Moreno and Umerez 2000). However, the appeal to inter-level causation in biological systems may oscillate between two different interpretations of the concept: on the one hand, the causal influence of an entity located at a given level of description on an external entity located at another (upper or lower) level of description; on the other hand, the causal influence of an entity, taken as a whole, on its *own* parts. As we will discuss, while it might seem quite obvious that closure involves inter-level causation in the first sense, a more difficult issue is whether this holds also for the second sense, which requires complying with more restrictive conceptual conditions.

Our argument in this chapter will be twofold. On the one hand, we will argue that closure can be consistently understood as an emergent regime of causation even though the autonomous perspective is interpreted, as we do, as being fundamentally committed to a *monism* (of properties). On the other hand, we will maintain that, although the mutual relations between constraints are such that the very existence of each of them depends on their being involved in the whole organisation, an emergent closed organisation does *not* necessarily imply inter-level causation, be it upward or downward, in the restrictive sense of a causal relation between the whole and its own parts (what we will label *nested* causation). Yet, as we will suggest, the appeal to inter-level causation in this sense (which is the philosophically more interesting and more widely discussed one) may possibly be relevant for organisational closure, if the adequate conceptual justification were provided.

The structure of the chapter being quite complex, let us provide a synthetic overview. In Sect. 2.1 we discuss one of the main philosophical challenges to the idea of emergence – Kim's exclusion argument – by focusing on the fact that it applies to a specific account of emergence formulated in terms of supervenience and irreducibility. In Sect. 2.2, we recall the distinction between two dimensions of the debate about emergence – ontological irreducibility and epistemological non-derivability – and clarify that a pertinent defence against the exclusion argument can be expressed, as we shall demonstrate, exclusively in terms of irreducibility. Section. 2.3 offers a conceptual justification of emergent properties and argues that configurations, because of the relatedness between their constituents, possess ontologically irreducible properties, providing them with distinctive causal powers. In Sect. 2.4, we focus on the specific case in which configurations exert distinctive causal powers as constraints, and argue that closure can be taken as a specific kind of higher-level emergent configuration (an organisation), ontologically irreducible and possessing distinctive causal powers: in particular, we will emphasise self-determination itself, which grounds most other features of autonomous systems.

Section. 2.5 concludes the analysis, and claims that closure can be justifiably taken as an emergent biological causal regime without admitting that it inherently involves inter-level causation in the precise sense of nested causation. Yet, the connection between closure and nested causation remains an open issue requiring further theoretical and scientific research.

2.1 The Philosophical Challenge to Emergence

The very idea of closure as a "distinctively biological" regime of causation cannot be justified unless it can be shown that, in some way, a given system possesses characteristic properties by virtue of which it may exert distinctive causal powers. A conceptual justification of emergence seems then to be a necessary requirement for a coherent account of biological causation.

Philosophical work on emergence began during the late-nineteenth and early-twentieth centuries, with the writings of the so-called "British Emergentists" (Mill 1843; Alexander 1920; Lloyd Morgan 1923; Broad 1925), and has developed considerably over recent decades.[1] As has often been underscored, a central contribution to this debate was made by Jaegwon Kim, who developed one of the most articulated conceptual challenges to the idea of emergence (Kim 1993, 1997, 1998, 2006).

In a recent survey of these issues (Kim 2006), Kim recalls what are, in his view, the two necessary ingredients of the idea of emergence, i.e. supervenience and irreducibility. *Supervenience* is a relationship by virtue of which the emergent property of a whole is determined by the properties of, and relations between, its realisers. As the author himself puts it (Kim 2006: 550):

> *Supervenience*: If property M emerges from properties $N_1, \ldots N_n$, then M supervenes on $N_1, \ldots N_n$. That is to say, systems that are alike in respect of basal conditions, $N_1, \ldots N_n$ must be alike in respect of their emergent properties.

In turn, *irreducibility*, and more precisely, according to Kim, *functional* irreducibility, is expressed as follows:

> *Irreducibility of emergents*: Property M is emergent from a set of properties, $N_1, \ldots N_n$, only if M is not functionally reducible with the set of the N_s as its realizer (Kim 2006: 555).

Given the account of emergent properties in terms of supervenience and irreducibility, the central issue is whether these properties may possess distinctive causal powers. In his work, Kim has developed several lines of criticisms vis-à-vis emergence. The one that is particularly relevant here claims that emergent properties are exposed to the threat of epiphenomenalism. Kim's argument on this matter is known as the "exclusion argument", and has been expounded on several occasions. Very briefly, the idea is the following. If an emergent property M emerges from some

[1] See (Sartenaer 2013) for an informed and insightful analysis of both the history and contemporary structure of the debate about emergence and reduction.

basal conditions P, and M is said to cause some effect, one may ask "why cannot P displace M as a cause of any putative effect of M?" (Kim 2006: 558). If M is nomologically sufficient for whatever effect X, and P is nomologically sufficient for M (because of the supervenience relation), it seems to follow that P is nomologically sufficient for X, and M is "otiose and dispensable as a cause" for X. As a result, invoking the causal power of emergent structures would be useless, since it would be epiphenomenal.

The exclusion argument has crucial implications for the debate about emergence and reduction. If one admits (1) that the relation between M and P is correctly described in terms of supervenience and (2) the validity of what we will call here the *principle of the inclusivity of levels,*[2] i.e. "the idea that higher levels are based on certain complicated subsets from the lower levels and do not violate lower level laws" (Emmeche et al. 2000: 19), then two problematic consequences follow.

First, the explanation is exposed to the danger of causal drainage. Indeed, if the causal powers of an emergent entity can be reduced to the causal powers of its constituents, and if, as may indeed be the case, there is no "rock-bottom" level of reality, then it seems that "causal powers would drain away into a bottomless pit, and there would not be any causation anywhere" (Campbell and Bickhard 2011: 14).[3] Second, if there were some scientifically justifiable rock-bottom level of reality (which is a far from trivial assumption),[4] and causal drainage were blocked, the exclusion argument would force reductive physicalism (see Vicente 2011 for a recent analysis). In this second case, any appeal to distinctively biological causal relations (such as closure itself, and related notions such as "integration", "control" and "regulation" etc.) would, at best, constitute an heuristic tool, unless it could be

[2]We take here the notion of "inclusivity of levels" as being analogous to Kim's "Causal Inheritance Principle" (Kim 1993: 326), according to which if a property M is realised when its physical realisation base P is instantiated, the causal powers of M are identical to the causal powers of P. By opting to use the term "inclusivity of levels", we wish to emphasise the idea that in the natural world, all causes are either physical or the result of the interaction between physical entities: no special causes (vitalist, spiritual, etc., that are not physically instantiated) are introduced at different levels, e.g. at the biological and mental ones. It should be noted that Kim's argument requires also the Causal Closure Principle as a premise, in the sense that the ultimate reduction of an emergent property to its fundamental realisation base is possible only if the basal level is causally closed (Kim 2003). Yet, we maintain that the validity of the inclusivity of levels does not necessarily require an appeal to the causal closure: emergent causal powers can be reduced to basal powers even though the latter are not shown or are supposed to be closed. Consequently, the argument that we develop in this chapter does not depend on the Causal Closure Principle.

[3]In Kim's intentions, the exclusion argument is originally targeted at mental causation and is not supposed to imply causal drainage. As a matter of fact, Kim himself has vehemently tried to avoid causal drainage as the ultimate consequence of the argument in favour of reduction. Moreover, on the basis of a commitment to the Standard Model and its lowest level of fundamental physical particles, he rejects the arguments based on the possibility of the absence of a rock-bottom level of reality. For a detailed discussion of these issues see, for example, Block's criticism of Kim's reduction argument (Block 2003) and Kim's reply (Kim 2003).

[4]The idea of a basic level with self-sufficient basic entities has been deeply questioned in microphysics, the very domain reductionist approaches appeal to as fundamental, where relational and heuristic accounts have been developed (Bitbol 2007).

demonstrated that said relations can be adequately reduced to physical causation or, more generally, to any "more fundamental" regime of causation.

In both cases, the very possibility of biological explanation is menaced. An adequate justification of a distinctive regime of biological causation should be provided, in order to (1) avoid the danger of endless causal drainage and (2) make the biological explanation theoretically independent from the physical and chemical ones, and directly related to the specificity of biological phenomenology, instead of being derived from lower level explanations and dependent on a single physical "theory of everything" (Laughlin and Pines 2000).

2.2 Irreducibility Versus Non-derivability

Before addressing the exclusion argument, let us make a preliminary conceptual distinction between two dimensions of the debate about emergence, i.e. irreducibility and non-derivability.

The exclusion argument challenges the status of emergent properties as causal agents of the world: how can a property be supervenient on something while being, at the same time, irreducible, and then possessing distinctive causal powers? An appropriate reply should then deal with the ontological issue of irreducibility, and justify emergent properties by showing that they are something ontologically "new" with respect to their realisers. Irreducibility, therefore, is inherently linked to *ontological novelty*.

Irreducibility should not be confused with the possible non-derivability of emergent properties from the emergence base, which is an *epistemological* issue. Non-derivability refers to the fact that given a description of the properties of the realisers, it is not possible to predict, explain or deduce the emergent properties of the whole.

As a matter of fact, most of the philosophical debate has tended to merge the two issues,[5] and to take both irreducibility and non-derivability as marks of emergence: emergent properties are not only irreducible but also, and crucially, non-derivable. Consider, for instance, the classic distinction between "resultant" and "emergent" properties, which is based precisely on criteria of non-derivability (or "non-deducibility", in Kim 2006: 552).[6] Resultant properties are aggregative properties, which the whole possesses at *values* that the parts do not (i.e. a kilogram of sand has a mass that none of its constituents has). Emergent properties, in turn, are properties of a *kind* that only the whole possesses, whereas the parts do not (i.e. a system can be alive, whereas none of its parts are alive). Although resultant

[5] The distinction has however been formulated, for instance, by (Silbersten and McGeever 1999), according to whom *epistemological* emergence concerns models or formalisms, while *ontological* emergence involves irreducible causal capacities. Here, we follow this conventional distinction.

[6] (Van Gulick 2001) refers to resultant and emergent properties as "specific value emergent" and "modest kind emergent" properties, respectively.

properties can be said to be, in a general sense, irreducible to the properties of their realisers, they are not what British emergentists (and most contemporary authors) had in mind when they talked about emergence. In fact, when appealing to notions like "unpredictability" or "unexplainability" as the mark of emergence, most authors are focusing on epistemological non-derivability.[7]

Yet we maintain that ontological irreducibility and epistemological non-derivability are logically distinct dimensions, and call for independent philosophical examinations. In what follows, we will discuss them separately, as two different issues.

On the one hand, we will develop throughout most of the chapter, a philosophical defence of emergence against the exclusion argument and the danger of epiphenomenalism, by relying exclusively on the irreducibility of emergent properties, without addressing the issue of their non-derivability. Emergent properties, we will argue, can be defined *exclusively in terms of irreducibility* and, crucially, they provide the system with distinctive causal powers *even though* they are derivable from their emergence base.

On the other hand, the issue of the non-derivability of emergent properties may play an important role in the discussion on whether emerging properties enable a system to exert inter-level causation between the whole and the parts. As we will suggest in the last part of the chapter (Sect. 2.5.2), if an emergent property is proven to be also non-derivable from the properties of the constituents, it may be possible to interpret the relationship between the whole and the parts as involving inter-level causation, because of the epistemological gap between them.

2.3 Irreducibility and Emergence

The aim of this section is to offer, in response to the exclusion argument, a conceptual justification of emergent properties provided with irreducible and distinctive causal powers. The core of the argument consists in suggesting that a coherent account of emergent causal powers can be obtained by rejecting the identification between the "supervenience base" and the "emergence base" of a property. As we will propose, a property of a whole can be functionally reducible to the set of properties of its constituents (its supervenience base) while being functionally irreducible to, and hence emergent on, various categories of entities that are distinct from that set. Once the distinction between the supervenience and emergence base is conceded, the resulting account of emergence, we will argue, eludes the exclusion argument and justifies the existence of distinct regimes of causation, even when maintaining the principle of the inclusivity of levels.

The argument will proceed in two steps. First, we will advocate (Sect. 2.3.1) an interpretation of the relation between the whole and the parts in terms of

[7]Crutchfield, for instance, distinguishes between two different definitions and classes of models of emergence, according to two different limitations in our capability "in principle" to describe emergent phenomena: nonpredictability and nondeducibility (Crutchfield 1994).

relational mereological supervenience, according to which a supervenience relation holds between the whole and the *configuration* of its own constituents, and not between the collection of constituents taken separately. We will then put forward a *constitutive* interpretation of relational supervenience, according to which supervenient properties can in principle be reduced to the configurational properties of the supervenience base. The main implication is that a supervenient property M and its basal properties $S_1, \dots S_n$ have identical causal powers. In the adoption of such a constitutive interpretation of relational supervenience lies, in our view, the *monist* stance of the autonomous perspective.

Second, we suggest (Sect. 2.3.2) that, even under the constitutive interpretation of relational mereological supervenience, a relation of emergence (as irreducibility) holds not between M and configurational properties, but instead between configurational properties and the properties of different categories of entities which do not belong to the configuration. Consequently, configurations can be justifiably said to possess irreducible and emergent properties and hence be able to exert non-epiphenomenal causal powers (in particular, as recalled in Sect. 2.4, as constraints) even under a monist interpretation of the autonomous perspective.

2.3.1 Supervenience and Constitution

The logic of the exclusion argument is based on the way in which the relation between an emergent property M of the whole W and the set of basal properties $N_1, \dots N_n$ of its constituents P is conceived. Namely, the relation is held to be simultaneously one of (mereological) supervenience and functional irreducibility, while assuming at the same time, as mentioned above, the validity of the principle of inclusivity of levels.

In his *Mind in a Physical World*, Kim paved the way for an answer to the exclusion argument capable of maintaining the inclusivity of levels, by clarifying the terms of the supervenience relation, and particularly specifying how the supervenience base is to be conceived. Kim argues that emergent properties are *micro*-based *macro* properties, i.e. second-order properties emerging out the first-order properties and relations of the basal constituents (Kim 1998: 85–86). The central idea is that the relevant supervenience base is not a set of properties of constituents taken individually or as a collection, but rather the properties of the *configuration* of constituents, i.e. the whole set of inherent *and* relational properties of the constituents. In other words, mereological supervenience should not be interpreted as atomistic but rather relational (see also Thompson 2007: 427–8; Vieira and El-Hani 2008).[8]

[8]The debate between a relational and an atomistic interpretation of the supervenience and emergence bases has a long history that dates back to the first formulations of the notion of Emergence in British Emergentism. In Alexander's framework, for example, space and time, the lower level on

The move to adopt relational mereological supervenience makes configurations of constituents the relevant supervenience base. The basal properties $S_1 \ldots S_n$ that bring about a supervenient property M are not the properties of the collection of constituents taken separately, but rather the configurational properties of the constituents *qua* constituents (including their mutual *relations*, which alter their intrinsic properties as separate elements), which appear only when the configuration is actually realised. If the basal constituents actually and collectively constitute a global pattern or system W, then their properties would now include those generated by their being involved in specific relations and interactions with other elements.

The adoption of relational mereological supervenience has relevant implications for the question concerning the distinctive causal powers of the supervenient property with respect to its supervenience base. Indeed, the idea that emergent properties would be reducible to the properties of the constituents taken in isolation seems to be excessively committed to an atomistic view of nature, which does not take relations into account (Campbell and Bickhard 2011). In turn, the claim that a supervenient property M is in principle reducible to the set of configurational (i.e. including relations) properties $S_1, \ldots S_n$ of its constituents is more convincing (again, by assuming the principle of inclusivity of levels), since configurations are far richer and more complex determinations than the mere collection of intrinsic properties of constituents.

Accordingly, we hold that relational supervenience does not imply functional irreducibility but rather, on the contrary, *constitution*: M supervenes on $S_1, \ldots S_n$ since it consists of $S_1, \ldots S_n$. A supervenient property M of a whole W corresponds to the set of configurational properties $S_1, \ldots S_n$ of its constituents (its supervenient base B). The set of the (relevant) configurational properties of the constituents of the system is, at least in principle, equivalent to the supervenient property. Hence, if M can be functionally reduced to the set $S_1, \ldots S_n$ of configurational properties of its constituents, it follows that it cannot possess distinctive causal powers[9] since, in fact, they are equivalent.[10]

---

which the whole natural world emerge, are relational concepts, not definable separately (Alexander 1920). The opposition between atomistic and relational approaches is particularly evident in Lloyd Morgan's work. He opposes to the billiard balls model of extrinsic interactions, the idea of relatedness based on inherent relations, which contributes to specifying the properties of the terms involved in the relation (Lloyd Morgan 1923: 19). It is also worth noting that, according to some authors, Kim's reference to relations is still made in a fundamentally atomistic framework, and does not imply a clear commitment to relational mereological supervenience, which implies the idea that relations "do not simply influence the parts, but supersede or subsume their independent existence in an irreducibly relational structure" (Thompson 2007: 428).

[9]The interpretation of relational mereological supervenience in terms of constitution is consistent, we argue, with the position developed by (Craver and Bechtel 2007) within their mechanistic framework. As they suggest, relations between constituents located at different levels in a mechanism are better understood as constitutive relations (pp. 554–555). See Sect. 2.5.1. below for a detailed discussion.

[10]For simplicity, we will only refer, from now on, to "configurational properties $S_1, \ldots S_n$" (equivalent to "property M"), and to "configuration C" (equivalent to "supervenience base B").

Yet, as we will claim in the following section, a coherent account of emergent properties provided with distinctive causal powers can still be provided, even under the constitutive interpretation of whole-parts relations.

2.3.2 *A Reply to the Exclusion Argument*

Our reply to the exclusion argument consists of arguing that even though supervenient properties (M) have no distinctive causal powers with respect to the configurational properties $S_1, \ldots S_n$ of the constituents, $S_1, \ldots S_n$ themselves (which are equivalent to M, because of constitution) are irreducible properties which may generate distinctive causal powers. Accordingly, $S_1, \ldots S_n$ can be said to be genuinely emergent. In other terms, there is an interpretation of emergence that is compatible with a monist stance.

Here is our argument. A given configuration C of elements of a whole W is identified by a set of (possibly dynamic) distinctive constitutive and relational properties $S_1, \ldots S_n$. On the basis of this set of distinctive properties, a configuration is functionally irreducible to any entity that does not *actually*[11] possess the same set of properties. We claim that a relation of emergence holds between a configuration C and any emergence base P whenever C is irreducible to P, i.e. if C possesses some distinctive set of configurational properties that P does not possess, such that C does *not* supervene on P. The reader would immediately note that this characterisation of emergence is very general, and could in principle include a wide range of obvious and uninteresting cases of P, which would not be considered salient for the philosophical debate on emergence. This is correct, and we deal with this issue just below. Yet, let us point out here that, as (Campbell and Bickhard 2011: 18; see also Teller 1986) have highlighted, appealing to configurations seems to be a sufficient answer to the danger of causal drainage and epiphenomenalism. The crucial point, as mentioned above, is that configurations include relational properties, which cannot be reduced to intrinsic properties, i.e. properties of constituents taken in isolation. Relatedness is ontological novelty. Consequently, because of relatedness (again: *actual* relatedness), configurations may possess distinct causal powers that would not otherwise exist.[12]

[11]It is important to emphasise that configurational properties must be actually realised, and not just "dispositional". As a consequence, a configuration C is functionally irreducible, in this account, also to those entities that would possess the "potential disposition" to actualise these properties.

[12]It might be objected that not *any* relational property gives rise to distinct causal powers. For instance, spatial relations do not seem to be relevant candidates in this respect while, for instance, relations that express energetic bonds among constituents do. In our view, useful specifications could indeed be offered on this point. Yet, we do not know whether a distinction between relevant and irrelevant classes of relational properties could (and should) be established *a priori*: hence, we do not provide further details in this book. For a related account of emergence, see also (Hooker 2004).

To avoid confusion, it is important to stress again that this account, in contrast with most existing ones, defines emergence exclusively in terms of ontological irreducibility, leaving aside the issue of the epistemological non-derivability of C from P. C is emergent on P if it possesses some set of *new* (relational) properties $S_1, \ldots S_n$ which P does not possess, and which are then irreducible to the set of properties $N_1, \ldots N_n$ of P. A different issue, which is irrelevant here, is whether one can derive or predict $S_1, \ldots S_n$ from $N_1, \ldots N_n$. In particular, $S_1, \ldots S_n$ would be irreducible *even if* they were derivable, because of the novelty introduced by the relations between constituents.

At this point, given the constitutive relation between the whole and its constituents advocated so far, one may wonder what exactly configurations do emerge on. Following our definition, three main kinds of emergent base P can be logically identified. Firstly, the configuration C is not supervenient, yet is emergent on the properties of any proper *subset* P_{sset} of its constituents (its parts). A wheel has emergent properties and distinctive causal powers on any subset of itself (e.g., a half-wheel). Secondly, the configuration C is not supervenient, yet is emergent on its *substrate* P_{sstr}, i.e. the collection of its constituents taken separately, as if they were not constituents (so to speak, the "potential ingredients" of a configuration). A wheel is emergent on the collection of molecules taken as if they were not actually assembled as a wheel.

Thirdly, and most importantly, the configuration C is not supervenient, yet is emergent on its *surroundings* P_{surr}, i.e. each set of external elements that does not actually constitute C. The wheel is emergent on each set of external molecules or entities, which are not actual constituents of it. In particular, given that a very broad set of entities might be included in P_{surr}, only relevant instances will actually be considered: in particular, the reference to surroundings P_{surr} will be restricted to those P_{surr} on which the configuration C has causal effects, by virtue of its emerging properties. As we will discuss in the following section, this is precisely the relevant case with regard to biological systems.

At this point, we have all the elements required to formulate our reply to the exclusion argument. The argument claims that emergent properties cannot be such unless it can be shown that they possess distinctive causal powers; at the same time, it seems that, as supervenient properties, they do not possess new causal powers with respect to their supervenience base. Hence, they are epiphenomenal. To this, we reply that emergent properties do not need to be irreducible to their supervenience base to possess distinctive causal powers: what matters is that configurations, because of relatedness, possess irreducible properties with respect to their subsets, substrate and (relevant) surroundings. Supervenience and emergence are then *alternative* notions: either a set of properties is supervenient on another one (in which case there is constitution between them), or it is emergent (in which case there is irreducibility).

Let us stress again that this way of conceiving emergence, interpreted exclusively as ontological irreducibility, is indeed very general. For instance, all chemical bonds are configurations emergent on their parts, substrate and surroundings, since they realise new relations, and therefore possess distinctive configurational properties.

Yet, the fact that this definition covers also irrelevant or obvious cases is, we argue, the price to pay for making it compatible with the monist stance of the autonomous perspective, represented by the constitutive interpretation of the relations between the whole and the parts. More generally, we hold that this characterisation of emergence is *sufficient* to provide a justification for the appeal to distinctive and irreducible causal powers in the scientific discourse (Laughlin et al. 2000), specifically in biology. Emergence appears whenever scientists are dealing with a system, such as a biological system, whose properties are irreducible to those of its isolated parts, substrate and surroundings. In such cases, one must introduce new observables, relations and causal powers, which exist only within that very system, and not in its emergent base.[13]

2.4 Constraints and Closure as Emergent Determinations

By virtue of their relatedness, configurations possess emergent properties and may exert distinctive causal powers on their surroundings that can take different forms, in accordance with the kind of systems under consideration. Let us focus here on the case in which these causal powers are exerted as constraints, which can in turn be organised as closure.

As discussed at length in Chap. 1, constraints are those configurations that, while exerting a causal action on a set of physicochemical processes and reactions (involving the movement, alteration, consumption, and/or production of entities, in conditions far from thermodynamic equilibrium), can also be shown, at the relevant time scale, to be conserved with respect to them.

By using the labels introduced in this chapter, we can rephrase the conditions (see Chap. 1, Sect. 1.3) under which an entity can be taken as a constraint as follows. Given a particular P_{surr}, a configuration C acts as a constraint C_{constr} if:

1. At the relevant time scale τ, C_{constr} *exerts a causal action on* P_{surr}, i.e. there is some observable difference between P_{surr} and P_{surr}^{c} (P_{surr}^{c} is P_{surr} under the causal influence of C_{constr} by virtue of relational properties $S_1, \ldots S_n$);
2. At τ, C_{constr} *is conserved throughout* P_{surr}, i.e. there is a set of emerging properties $S_1, \ldots S_n$ of C which remain unaffected throughout P_{surr}.

In Chap. 1 we claimed that constraints constitute a distinct regime of causation because, by fitting these conditions, they are irreducible to the thermodynamic flow on which they exert a causal action. In particular, for a given effect, one can make a

[13]It is worth noting that the relation between the emergent properties and their emergence base can be interpreted both synchronically and diachronically. Being based on novelty, in fact, the irreducibility to any entity that does not belong to an actual configuration is in principle compatible with both dimensions of emergence.

conceptual distinction between two kinds of causes: the "material" ones (inputs or reactants), which do not meet condition 2, and constraints.

Now, this characterisation of constraints relies precisely on a justification of the emergent nature of those relevant properties in virtue of which they satisfy the above conditions. And – we submit – our conception of emergence provides such justification: constraints are irreducible to the thermodynamic flow insofar as they possess properties that emerge from that flow, because of the relatedness of their configurations. In other words, to explain why constraints are irreducible to the relevant P_{surr} on which they act (and then why the exclusion argument does not apply to them) one has to appeal to the ontological novelty of their emergent properties with respect to P_{surr}, a novelty generated by the relatedness of their configurations at the relevant time scale. Similarly, their emergent configurational properties support the distinctive causal powers of constraints: at biological relevant time scales, constraints – as discussed in Chap. 1 – are enabling, in the sense of causally contributing to the maintenance of the whole organisation, which would otherwise be a highly improbable (or *virtually* impossible) phenomenon.[14]

It is worth noting that, in line with our characterisation, constraints can be said to emerge on a wide spectrum of entities belonging to P_{surr}, which goes far beyond the case of processes and reactions in far-from-equilibrium thermodynamic conditions. Yet, as highlighted earlier, we restrict our analysis to the subset of P_{surr} on which constraints exert a causal action, because this is the case in which the (ir)reducibility of one regime of causation to another has explanatory relevance for biological systems.

Let us now turn to closure. As discussed in Chap. 1, closure generates an organisation in which the network of constraints achieves self-determination as collective self-constraint. In the terms of this chapter, hence, closed organisations are then *a specific kind of higher-level configurations* C_{org}, whose distinctive feature consists in the fact that their constituents are themselves configurations C_{constr} acting as constraints. As such, we claim that closure, because of the relatedness between the constraints, generates itself ontological novelty and new emergent properties, possibly supporting distinctive causal powers of the organisation. In particular, let us distinguish three aspects.

First, as we already mentioned in Chap. 1, closure inherits the irreducibility of constraints to the surroundings P_{surr}, the thermodynamic flow. To the extent that no adequate account of constraints can be provided by reducing them to the causal regime of thermodynamic changes, *a fortiori*, closure of constraints cannot itself be reduced to a closed network of processes and changes. Hence, a description of biological systems in terms of pure thermodynamics would not be able to account for their organisation.

[14] In Chap. 1 we argued that, in most cases, C_{constr} does not extend the set of possible behaviours of P_{surr}, i.e. P_{surr} could in principle (although it is highly unlikely) exhibit, at different time scales, the behaviour of $P_{surr}{}^{c}$ without the action of C_{constr}.

Second, and crucially, organisations themselves possess additional emergent properties with respect to their substrate P_{sstr}, namely the collection of their constitutive constraints taken separately. When constraints actually realise closure, their relatedness generates new emergent properties that none of them would possess separately. One of these properties is, of course, self-determination itself: individual constraints cannot determine (or, more precisely, maintain) themselves, only their collective organisation can.[15] Hence, the capacity to self-determine is, in the biological domain, an emergent property generated by the specific relatedness of the closure of constraints. As mentioned in the Introduction, the whole autonomous perspective can be seen as an exploration of the distinctive features of biological organisms generated by their emergent capacity of self-determination. In a way, each chapter of this book focuses on a set of properties or capacities stemming from the realisation of organisational closure: this holds for agency, biological complexity, multicellular organisations, cognition, and so on .... The philosophical justification of the irreducible ontological novelty of closure, through the appeal to the relatedness between the constitutive constraints, is therefore a pivotal step toward the elaboration of the whole theoretical framework.

To make this point clear, let us mention a specific example, which introduces the following Chap. 3. As we will discuss at length, closure generates functionality. As has been recently argued (Mossio et al. 2009b; Saborido et al. 2011), when they are subject to closure, constraints correspond to biological functions: performing a function, in this view, is equivalent to exerting a constraining action on an underlying process or reaction in an organised system. All kinds of biological structures and traits to which functions can be ascribed satisfy the above definition of constraint, although at very different temporal and spatial scales. Some intuitive examples in addition to the vascular system mentioned above include, at different scales: enzymes (which constrain reactions), membrane pumps and channels (which constrain the flow of ions through the membrane) and organs (such as the heart which constrains the flow of blood), among others. The emergence of closure is then the emergence of functionality within biological organisation: constraints do not exert functions when taken in isolation, but only insofar as they are subject to a closed organisation. As a consequence, the defence of a naturalised account of functionality as a distinctive biological dimension (developed in Chap. 3) fundamentally relies on the justification of the emergent and irreducible nature of closure advocated here.

Third, closure is specifically defined with respect to the emergence base P_{surr}, constituted by a set of processes and changes occurring in conditions far from thermodynamic equilibrium. The two causal regimes, although mutually irreducible, realise a two-way interaction in biological systems, to the extent that constraints act on thermodynamic processes and changes, which in turn contribute to reproducing or maintaining these constraints. Hence, it might be tempting to

[15] Whereas, as we mentioned in Chap. 1, Sect. 1.4 above, individual self-determination does in fact occur in physics, in the case of self-organising dissipative structures.

conclude that closure (just like any form of self-maintenance) inherently involves not just emergent causation, but also inter-level causation, at work between the two causal regimes. Yet there are several reasons to resist this temptation, at least insofar as particularly controversial kinds of inter-level causation are concerned.

2.5 Inter-level Causation

The issue of inter-level (be it upward or downward) causation has been, explicitly and otherwise, a central aspect of the debate about emergence from its earliest beginnings,[16] since the very concept of emergence carries on the issue of the relations between properties at different levels.

In Kim's account, attributing causal powers to emergent properties necessarily implies downward causation. Let us recall his argument that, as we discussed, identifies the supervenience and the emergence bases. Let M and M* be two emerging properties, and suppose that M causes M* (a case of "same-level" causation). As an emergent property, M* has an emergence base, say P*. Given the supervenience relation, P* is necessary and sufficient for M*: if P* is present at a given time, then M* is also present. Accordingly, it is unclear in what sense M could play a causal role in bringing about M*: given P*, its role would be useless, unless M is in fact somehow involved in causing P*. In other words, the same-level causation of an emergent property makes sense only if this implies the causation of the "appropriate basal conditions from which it will emerge" (Kim 2006: 558). Consequently, causation produced by emergent properties seems to imply, in all cases, downward causation in the sense of a causal influence exerted by an emergent property on the basal conditions of another emergent property.

Yet, as Kim himself has argued (Kim 2010), this general form of downward causation, i.e. a causal influence exerted by an entity at a higher level on a different entity located at a lower level, is indeed widespread and unproblematic. In particular, this interpretation of downward causation applies straightforwardly to self-maintenance and closure, which inherently involve, as discussed above, upward and downward causation between constraints and dynamics, with each being located at different levels of description.

---

[16]According to Lloyd Morgan, "[...] when some new kind of relatedness is supervenient (say at the level of life), the way in which the physical events which are involved run their course is different in virtue of its presence-different from what it would have been if life had been absent. [...]. I shall say that this new manner in which lower events happen – this touch of novelty in evolutionary *advance depends on* the new kind of relatedness" (Lloyd Morgan 1923: 16). According to (Stephan 1992) Lloyd Morgan's passage could admit different interpretations, such as that of a logical claim about supervenience. On the contrary, McLaughlin asserts: "In Morgan one finds the notion of downward causation clearly and forcefully articulated" (McLaughlin 1992: 68).

The more controversial form of downward causation would be that exerted by a whole on its own constituents (in Kim's terms, "reflexive" downward causation, Kim 2010: 33). According to (Emmeche et al. 2000), there are various possible interpretations of reflexive upward and downward causation. In their view, the only non-contradictory versions of the concept are those that interpret downward causation in terms of "formal" causation (Emmeche et al. 2000: 31–32), such that the whole exerts *a constraining action on its own constituents*, by selecting specific behaviours from among a set of possible ones. This interpretation can be taken, as Emmeche and his co-authors claim, as the standard and possibly more compelling one of downward causation, and it is very close to the original proposal by (Campbell 1974).[17]

As an illustration, consider Sperry's classic example of the wheel rolling downhill (Sperry 1969). On the one hand, the various molecules generate the wheel as a whole, and on the other, as (Emmeche et al. 2000: 24) explain:

> none of the single molecules constituting the wheel or gravity's pull on them are sufficient to explain the rolling movement. To explain this one must recur to the higher level at which the form of the wheel becomes conceivable.

The set of configurational properties of molecules is assumed here to underdetermine their behaviour so that, in order to explain it, one needs to appeal to a property of the whole (in this case: the form of the wheel) that would generate a causal influence (a selective constraint) exerted on its own constituents.

Because of the (assumed) under-determination of constituents by configurational (intrinsic and relational) properties, constituents' behaviour is partly determined, in a functionally irreducible way, by the whole to which they belong. In particular, this train of thought seems to apply equally to biological systems, in which the behaviour and dynamics of the parts appear to be, in an important sense, determined (notably through regulation and control functions) by the downward causation exerted by the whole system to which they belong.

In what follows, we will examine whether self-maintenance and closure do indeed involve some form of reflexive inter-level causation, intended as a particular form of constraint exerted by the whole on its parts. As we will argue (Sect. 2.5.1), there seems to be no compelling argument in favour of a positive answer within our framework, at least under the monist assumptions adopted so far. Alternative conclusions could be obtained (Sect. 2.5.2) by rejecting some of these assumptions, or by shifting the analysis to an epistemological or heuristic dimension.

Before continuing, a terminological clarification: A possible objection might contend that this debate somehow forces a narrow understanding of inter-level causation in terms of reflexive whole-parts causal influence, whereas the usual meaning in the biological domain refers to the non-reflexive case, where higher-

---

[17]Campbell defines downward causation as follows: "all processes at the lower level of a hierarchy are restrained by and act in conformity to the laws of the higher level" (Campbell 1974: 180). More recently, (Vieira and El-Hani 2008) have proposed a similar view, although they refer to "formal determination" instead of formal causation.

level entities interact with lower-level entities, the latter not being constituents of the former. Indeed, this interpretation of inter-level causation applies straightforwardly to biological organisation, and is inherently involved in the very notion of closure. In this sense, biological discourse requires a general concept of inter-level causation. To avoid ambiguities, we propose using different terms to refer to the two ideas: in what follows "inter-level causation" will be therefore used for the general non-reflexive case, and "nested causation" for the reflexive whole-parts case. This way, biological descriptions would be able to refer to inter-level causation, while at the same time avoiding incongruities with philosophical analyses.

2.5.1 *Why We Do Not Need Nested Causation in Biology*

The account of emergence and supervenience developed so far has relevant implications for the conception of nested causation.[18]

Concerning the supervenience base B – insofar as the principle of inclusivity of levels is maintained (but see Sect. 2.5.2 below), and the relation between an emergent property M and the configurational properties $S_1, \ldots S_n$ of B is conceived as constitutive – the exclusion argument applies more cogently to relational supervenience than to its atomistic version. Consequently, as (Craver and Bechtel 2007) emphasise, no nested causation can exist between an emergent property and its own supervenience base: there is no justification for claiming either that $S_1, \ldots S_n$ "generate" or "produce" M, or that M exerts downward causation on $S_1, \ldots S_n$. In particular, the closed organisation C_{org} does not exert causation on the whole network of constitutive constraints, and the whole network of constitutive constraints does not produce the closed organisation. Under the monist stance adopted so far, there is therefore no room for nested causation in the autonomous perspective.

Let us now consider the emergence base P of C, and its different versions discussed in Sect. 2.4. Is there nested causation between the organisation C_{org} and any *subset* P_{sset} of its constituents? In our view, by assuming the principle of the inclusivity of levels, the answer is no, since the properties of each P_{sset}

[18] In the philosophical literature, nested causation comes in two variants, synchronic and diachronic (Kim 2010: 34–36). On the one hand, *synchronic* nested causation refers to the situation in which upward and downward causation would occur simultaneously. In more technical terms, a supervenient property M acts causally on its supervenient base $S_1 \ldots S_n$ *at the same time* as the supervenience base generates M. On the other hand, *diachronic* (or diagonal) nested causation refers to the situation in which M acts on its own supervenience base $S_1, \ldots S_n$, causing its modification, but only at a subsequent time with respect to the upward determination. In this chapter, however, we assume that the distinction is irrelevant, since we question the very idea of the causal influence of M on $S_1, \ldots S_n$, be it synchronic or diachronic. In particular, in line with Sartenaer's detailed argument (Sartenaer 2013: 240–250), we assume in this chapter that all cases of diachronic nested causation are also synchronic and vice-versa.

(which may refer, for instance, to each individual constraint C_{constr}) are by definition configurational, so that the appeal to some constraint exerted by the whole would be redundant: configurational properties are so precisely because an entity belongs to a whole. To put it more straightforwardly, local constraints are not so "because they are under the causal influence of the whole". Also, no nested causation occurs between the whole and its *substrate* P_{sstr} because, in our account, the collection of its constituents taken separately (without their configurational properties) is an abstract description that does not correspond to the way in which constituents are organised in the system. Since, in the system, there is no such thing as a collection of unrelated constituents, they cannot be, *a fortiori*, involved in nested causation, or indeed any causation at all.

The case of the third kind of emergence base, the *surroundings* P_{surr}, is somewhat different. As discussed in Sects. 2.3 and 2.4, emergent configurations do exert a causal action on their surroundings, notably in the form of constraints. Yet, surroundings are by definition external to the configuration, which means that the constraints exerted by C_{org} on P_{surr} *can by no means be interpreted in terms of nested causation.*

The claim according to which constraints, in our framework, do exert causal powers, but not in the form of nested causation, has crucial consequences for the interpretation of self-maintenance and closure.

In the case of physical self-maintaining systems, the fact that the emergent configuration acts to maintain itself does not appear to constitute, *per se*, a case of nested causation, since the constraining action is exerted on the surroundings of the configuration, not on its own constituents. Let us again take the example of Bénard cells. An interpretation appealing to nested causation would claim that each cell (i.e. the emergent configuration) exerts a constraint on its own microscopic constituents, in the sense that the fact of belonging to a given cell *determines* whether a molecule rotates in a clockwise or anticlockwise direction. As (Juarrero 2009: 85) puts it:

> Once each water molecule is captured in the dynamics of a rolling hexagonal Bénard cell it is no longer related to the other molecules just externally; its behaviour is contextually constrained by the global structure which it constitutes and into which it is caught up. That is, its behaviour is what it is *in virtue of the individual water molecules' participation in a global structure.*

Yet, what we call the cell *is* the configuration of constituents, so that, as argued above, it is redundant to appeal to the whole set of constituents and relations to explain the behaviour of each constituent, whose characterisation already includes its relational properties as part of the configuration. Once a given molecule has been "captured" by the cell and has begun to rotate with the others, in what sense would it still be "constrained by the global structure"[19]? Let us have a closer look at this

[19] A satisfactory analysis of nested causation requires, then, a careful distinction between two ideas. One is the idea that a configuration is made up by a set of constituents, which have causal interactions between them. Explaining why a given molecule of water is rotating in a given manner at a given moment requires an appeal to its causal interactions with other constituents. And the

situation, the temporal steps being very important. At a given moment a macroscopic structure is formed. Once this structure is formed, it acts on the surrounding molecules, constraining their microscopic trajectories in such a way that they generate a thermodynamic flow that, in turn, contributes to the maintenance of the (otherwise decaying) macroscopic structure. The result is a causal regime in which the dissipative structure acts on its surroundings that, through this action, contribute to the maintenance of the very structure. Only in this loose sense the surrounding constrained processes could be said to being "parts" of the self-maintaining regime; yet, the causal action of the dissipative structure in each moment does *not* operate on its own constituents.[20]

Two reasons may explain why self-maintaining systems seem to be a case of nested causation. First, the description of the configurational properties of dissipative structures, which are available at a given moment, usually under-determines their behaviour. This is of course a crucial point; still, as discussed earlier, this should not be taken as sufficient reason for appealing to nested causal relations since, as pointed out in Sect. 2.3, it confuses epistemological non-derivability with ontological irreducibility (but see Sect. 2.5.2 below). Second, self-maintaining systems would not exist if they did not generate a causal loop between the whole configuration and its constituents. Yet, the crucial point is that, in our view, this loop is not a *direct* loop, but rather an *indirect and diachronic* one, realised through the action of the constraint on its surroundings. What might appear as an action exerted on the constituents is in fact exerted on the *boundary conditions* of these constituents.

In the light of these considerations, in particular, we do not think that the appeal to the supposed constraint exerted by the configuration on its own constituents in terms of *formal* causation is explanatory (again, under the monist assumptions adopted so far). The formal causation of the whole on its constituents would be in principle reducible to the constraining action exerted on the boundary conditions of these constituents, without loss of information or explanatory power.

---

reason why a set of constituents may exert a causal influence on other constituents is, of course, that all of them belong to the same system. The other idea, in contrast, is that the "whole system", including any specific constituent, would have a causal effect *on that very constituent*.

[20]This argument also applies to the relation between the whole and its parts in self-assembling structures. Let us take the example of protein folding. Protein folding is a process in which the parts – aminoacids – form small secondary (metastable) structures, which, once constituted, harness the surrounding interactions leading to the formation of new structures, and so on, till the global tertiary and quaternary folded structure is achieved. The whole folding process is a succession of formation of local wholes, acting as constraints on surrounding dynamical pieces that, later, will become wholes harnessing other pieces, and so on. Once the folding is achieved the whole has attained a relative thermodynamic equilibrium (it is a conservative structure) and it does not make sense to say that it acts constraining its parts. So, when we consider the temporal (diachronic) process of folding, the (local) "wholes" act on their surrounding, not on their constitutive parts; similarly, when we consider the protein already folded (synchronically) the whole does not act on its own parts either.

Let us now examine closure. Is there a characteristic aspect of closure that would justify, in contrast to simple self-maintenance, the claim according to which it realises nested causation?

The main difference between physical self-maintenance and closure is that, in the second case, self-maintenance is realised collectively by a network of mutually dependent constraints. In real biological systems, closure is realised through a very complex organisation of constraints, such that, in most cases, a given constraint exerts its action on surroundings that have already been subject to the causal influence of at least one other constraint. For instance, most enzymes act on reactions whose reactants are the result of the joint action of other constraints, including the membrane (through its channels and pumps). In these cases, it can be said that constraints act on entities that are already "within" the system, at least in the sense of having already been constrained by the system. This seems to be a clear difference with respect to simple self-maintaining systems, and one may then conclude that the closed organisation does act on its own constituents, and realises nested causation.

Yet we hold that this conclusion is incorrect, since it interprets those constrained processes and reactions as constituents of the organisation (which, we recall, is defined as a specific kind of higher-level configurations, as a closed configuration of mutually dependent constraints), whereas they are not. In biological systems, the constituents of the organisation are the constraints themselves, which realise collective self-maintenance. According to the constitutive interpretation of the relation between the whole and its constituents, the organisation as such does not possess emergent and distinctive causal powers with respect to the closed network of constraints which, in turn, exert causal powers on surroundings which are not themselves constituents of the network (although they are usually within the spatial borders of the system).[21] Accordingly, to use the terminological distinction introduced above, we maintain that closure does involve inter-level causation, but *not* nested causation.

A second reason why closure seems to inherently imply nested causation is that evoked by Kant (1790/1987), i.e. the fact that the existence of the constituents (the constraints) "depends on the whole". Indeed, the mutual dependence between constraints constitutes a fundamental difference between organisations and other configurations. In the second case the existence and maintenance of the constituents might not depend on their being involved in the configuration: one can decompose a wheel into its molecular elements, which would continue to exist as separate elements. The same holds for the microscopic constituents of a dissipative system. In contrast, closed organisations imply a more generative kind of relation between

[21]The physical processes on which the network exerts (constraining) causal powers can, in some cases, become members of the network itself, when they enter into configurations which act as constraints. Nonetheless, the network would exert causal powers on them for as long as they remained part of its surroundings, and would cease acting causally on them as soon as they started playing the role of constraints.

constituents (the constraints themselves), which exist only insofar as they are involved in the whole organisation. Actually, the appeal to formal causation advocated by several authors is essentially aimed, in our view, at capturing this distinctive feature of biological organisms.

Yet, these specific features of organisations do not require an ascription of distinctive causal powers to the whole, since closure can be realised through the network of mutual, usually hierarchical, causal interactions. Hence, "depending on the whole" could simply mean "depending on the whole network of interactions" without appealing to the whole as a causal agent emergent on its own supervenience base.

This interpretation of the whole-parts relation in biological organisation is particularly relevant because it applies to all those cases in which biological literature typically appeals to nested causation, i.e. all kinds of regulation and control mechanisms (see Chap. 1, Sect. 1.8 above) thanks to which organisms are able to (adaptively) compensate for internal and/or external perturbations (Piaget 1967; Fell 1997). What is frequently described as a causal action of the whole system on its own constituents, is in fact the result of the interaction among organised constraints (or subsystems of constraints) which can result, for instance, in the acceleration of the heart rate and glucose metabolism when the organism starts playing tennis (see Craver and Bechtel 2007: 559, for a detailed description of this example, and other relevant ones). In particular, inter-level control can be generally understood in terms of causal interactions among constraints located at different hierarchical levels of emergent organisation. In turn, as we claimed in Chap. 1, Sect. 1.8, regulation specifically concerns interactions among constraints at different orders of closure. Although both cases inherently require, as all biological functions do, the realisation of closure as well as inter-level (and inter-order) causation between hierarchically organised constraints, they do not involve nested causation exerted by the whole organism.

2.5.2 Why We Might, After All, Need Nested Causation in Biology

The rejection of nested causation depends on the constitutive interpretation of the supervenience relation adopted so far. It is an implication of our monist interpretation of the autonomous perspective. Indeed, the central goal of the analysis was to suggest that closure can be justifiably taken as an emergent and distinctively biological regime of causation *even* under a constitutive interpretation of supervenience. Yet, several strategies could be adopted to justify nested causation, and they might be successful and operational in some cases, including the biological domain, which is specifically under study here. To date, however, we believe that these strategies lack any compelling argument in favour of their adoption in Biology; their relevance is still under conceptual and scientific scrutiny. That is why, in our view, the constitutive interpretation of the whole-parts relation is still the wiser one. Let us discuss these strategies.

The first strategy is ontological and advocates that a non-constitutive interpretation of relational supervenience should be adopted, in order to admit causation of the whole on the constituents. In this interpretation, emergent properties can be simultaneously supervenient on *and* irreducible to configurations. For this ontological stance to be coherent, one must accept the violation of the inclusivity of levels, hence accepting the idea that the very same entity (say: a constituent of a configuration) may possess different properties, and then obey different laws or principles, at different levels of description. In other terms, it consists in rejecting the monist stance advocated so far. For instance, each molecule constituting the wheel would have the property to behave in a given way when considering the whole configuration, but would *not* possess the same property when looked at individually. Even though we are looking at the very same molecules under the very same conditions, their properties would vary according to the level of description, since the relevant laws and principle would also vary.

In our view, rejecting the principle of the inclusivity of levels could indeed be an important tool for adequately accounting for natural phenomena that would therefore require an appeal to nested causation. We have no principled objections to this position. Yet, we maintain that its relevance for the biological domain is still uncertain. As (Craver and Bechtel 2007) have convincingly argued, many (or most) apparent biological examples of nested causation (in particular cases of downward regulation) seem to be adequately explainable by appealing to what they call "hybrid accounts" involving intra-level causal interactions *between* constituents and inter-level constitutive relations, or to what we dubbed inter-level (not nested) relations. In those cases, an advocate of the constitutive interpretation of mereological supervenience could argue that the appeal to nested causation seems precisely to stem from an inadequate understanding of the role of configurations: the behaviour of the constituents appears to be influenced by the whole because the description focuses only on the internal properties of the constituents, neglecting the relational ones.[22] In a word, there seems to be no clear case in the biological domain for

[22]In the case of the wheel, for instance, one may say that if we describe a given molecule as a constituent of a wheel, we are already including in the description all constitutive and relational properties, which make it a constituent ("being in such and such position", "having such and such interactions and links with neighbouring molecules" etc.), and which determine its behaviour under specific conditions. For instance, a force (i.e. gravity) applied to a part will generate a chain of causal interactions between the constituents that, because of their individual configurational properties, will behave in a specific way. We will then call the collective pattern the "rolling movement of the wheel". Each molecule of the wheel will move in a specific way because its configurational properties force it to do so, and a complete description of the configurational properties of the individual constituent will suffice to explain why it behaves as it does. The fact that the constituents collectively constitute a wheel, whose macroscopic behaviour can be described as a rolling movement, does not add anything to the explanation of the individual behaviour. There are indeed causal interactions here, but no inter-level causation.

which the appeal to nested causation is mandatory. To a first approximation, self-maintenance and closure are no exceptions in this respect.

The second strategy is epistemological, and consists of justifying nested causation by demonstrating that it would be impossible, *in principle*, to determine the behaviour of a system through a description of its configurational properties. On the basis of such a negative result, the appeal to nested causation would be justified in epistemological terms, since there would be, in principle, no alternative description.[23] Yet, while arguments of "inaccessibility" have already been formulated in physics (Silbersten and McGeever 1999), and might for instance be relevant for describing dissipative structures, this is not the case in biology.[24] Consequently, there seems, to date, to be no compelling epistemological argument for admitting nested causation for biological systems.

The third strategy is heuristic. As a matter of fact, there are many cases, especially in complex systems, in which the available description of the configurational properties is insufficient to adequately determine the behaviour of the whole system. In these cases, which are indeed widespread, it might be useful to appeal to the configuration as a whole *as if,* by virtue of its emergent properties, it were exerting nested causation on its constituents, so as to provide a description capable of sufficiently determining the behaviour of the system. Since it is not committed to a theoretical non-constitutive interpretation of supervenience, the heuristic appeal to nested causation can be adopted as a pragmatic tool even within a constitutive interpretation of supervenience. Yet, such a heuristic use of nested causation would not point to any specific feature of the causal regime at work in biology (which is the object of this chapter), but would simply correspond to a scientific practice common to several scientific domains. In particular, as mentioned above, the strategy can be adopted for self-maintaining, closed systems for which, mostly because of their internal complexity, complete descriptions of their organisation are difficult to elaborate.

2.6 Conclusions

In order for closure to be a legitimate scientific concept rather than merely an epistemic shortcut, philosophical arguments must be provided in favour of its emergent and irreducible nature with respect to the causal regimes at work in other classes of natural systems. To do this, we developed a twofold argument.

---

[23]See (Bich 2012) for an epistemological discussion of the relationship between emergence and downward causation.

[24]It should be noted, however, that the issue is currently being explored by several biologists and theoreticians. For instance, a relevant proposal in this direction has recently been developed by Soto et al. (2008).

On the one hand, we argued that constraints are configurations that, by virtue of the relations existing between their own constituents, possess emergent properties enabling them to exert distinctive causal powers on their surroundings, and specifically on thermodynamic processes and reactions. When a set of constraints realises closure, the resulting organisation constitutes a kind of second-level emergent regime of causation, possessing irreducible properties and causal powers: in particular, organisations are able to self-determine (and more precisely to self-maintain) as a whole (something which none of their constitutive constraints are able to do). As we will see in the following chapters, most of the distinctive features of autonomous systems specifically rely on closure and organisation, which therefore play, as an emergent causal regime, a pivotal role in the autonomous perspective.

On the other hand, we advocated the idea that a coherent defence of closure as an emergent and irreducible causal regime does not need to invoke nested causation. Closed organisations can be understood in terms of causal (possibly inter-level) interactions between mutually dependent (sets of) constraints, without implying upward or downward nested causal actions between the whole and its parts. Biological emergence, accordingly, is logically distinct from nested causation, and one can advocate the former without being committed to the latter.

Again, we by no means wish to exclude the possibility that biological organisation might involve nested causation. As discussed earlier, various strategies could be adopted to adequately justify this idea, and promising explorations are currently underway. Nevertheless, we believe that these attempts are, as yet, incomplete, and do not offer compelling arguments in the biological domain. That is why we argue that biological organisation can be shown to be emergent and irreducible *even though* nested causation is, by hypothesis, ruled out.

3
Teleology, Normativity and Functionality

According to the autonomous perspective, the constitutive organisation of biological systems realises an emergent regime of causation, which we labelled *closure of constraints*. One of the crucial implications of the realisation of closure is that, as we will argue in this chapter, it provides an adequate grounding for a distinctive feature of biological systems, namely their *functionality*.

The concept of function is widespread in the language of all life sciences. At the scale of individual organisms, functions are usually ascribed to a variety of structures, traits, or processes that constitute the whole, such as, for instance, systems, organs, cells, and molecules. Similarly, functions are invoked when considering larger scales, so that organisms themselves, as well as populations and species, may be the object of functional ascriptions. Moreover, as Gayon points out (Gayon 2006: 480), functional ascriptions mostly tend to have a nested structure: parts of a functional entity can also perform functions and, reciprocally, systems containing functional entities may also be described as functional.

What is the status of the concept of biological function? At first glance there seems to be a broad consensus regarding the idea that functions play a genuine explanatory role in biology and the other life sciences: functional ascriptions are by no means simple descriptions of a trait, but rather provide an understanding of some of its essential properties and activities. To be sure, the explanatory role of functions seems to be so fundamental in life sciences that one could argue that biological explanations are *essentially* functional: in contrast to those at work in, for instance, physics or chemistry, biological explanations would be specific in this, i.e. in that they appeal to functions.

Many of the ideas developed in this chapter, as well as several portions of the text, come from Mossio et al. (2009b), and Saborido et al. (2011). Section 3.3.2 is partially based on Saborido's account of malfunction, exposed in his PhD dissertation (Saborido 2012) and in Saborido et al. (2014).

A. Moreno, M. Mossio, *Biological Autonomy*, History, Philosophy and Theory of the Life Sciences 12, DOI 10.1007/978-94-017-9837-2_3

Even though it is not our aim to adopt a final position in relation to this last issue, it cannot be denied that the concept of function is at the very heart of scientific discourse in life sciences. Yet it generates a major epistemological problem, since it seems to be, at least at first sight, at odds with the ordinary structure of scientific explanation, because of its characteristic dimensions, i.e. its *teleology* and *normativity*. But what does this actually mean?

On the one hand, functions have an explanatory role in accounting for the existence of function bearers. Affirming that (to cite a classic example) "the function of the heart is to pump blood" does not correspond to a simple description of what the heart does; rather, in addition, it means that this effect has specific relevance in explaining the existence, structure and morphology of hearts (see also Buller 1999: 1–7). Hearts exist to some extent because they pump blood. Functional attributions thus introduce a teleological dimension into the structure of explanation, in the sense that the existence of a trait could be explained by appealing to some specific effects or consequences of its own activity, which reverses the conventional order between causes and effects.

On the other hand, the concept of function possesses a normative dimension, to the extent that it refers to some effect that the trait is *supposed* to produce (Hardcastle 2002: 144). Attributing functions to a trait implies a reference to some specific norm, against which the activity of the trait can be evaluated. The claim that "the function of the heart is to pump blood" implies also that the heart must pump blood. Whereas, usually, causal effects simply occur, functional causal effects must occur.

Because of its teleological and normative dimensions, the concept of function seems then to be in conflict with the accepted structure of scientific explanation. The central question is then: is the concept of function a legitimate and admissible scientific concept?

To answer this question, two alternative strategies are possible. The first is an eliminativist one, and consists of denying that functions do in fact play an explanatory role. All functional claims can be reformulated in terms of an ordinary causal claim, without losing information or meaning. In this case functions would constitute, at best, a linguistic shortcut. The second strategy, in contrast, claims that while functional statements cannot be reduced to ordinary causal ones, they are compatible with the structure of scientific discourse. In this case, a naturalisation of teleological and normative dimensions is required, i.e. a justification of the idea that these dimensions are grounded in some objective features and properties of biological systems and, consequently, can be analysed in adequate scientific terms.

In this chapter, we will suggest that the autonomous perspective adopts the second strategy, and puts forward a naturalised "organisational" account[1] of functionality, based on the emergent properties of closure.

[1]For terminological clarity, note that we will dub "organisational account" (OA) the account of functions stemming from the view of living beings as organisationally closed systems, and, in particular, from the autonomous perspective.

3.1 The Philosophical Debate

Broadly speaking, the philosophical analysis of the concept of function is very old, to the extent that it has always developed hand in hand with scientific research into biological phenomena. However, the debate on functions, in its contemporary form, has been framed during the last four decades, during which an increasing number of studies have been conducted in philosophy of science and philosophy of biology (several collections have published that survey the recent philosophical debate: see Ariew et al. 2002; Buller 1999; Allen et al. 1998; Gayon and de Ricqlès 2010).

The contributions that gave rise to the contemporary debate were formulated during the sixties by Nagel (1961, 1977) and Hempel (1965) who, by adopting an eliminativist stance, tried to reduce functional statements to the nomological-deductive model (Hempel and Oppenheim 1948). Because of the difficulties inherent in their approach (Saborido 2012: 51–59), the vast majority of subsequent literature has focused on justifying functional discourse through naturalisation.

Current philosophical accounts of functions are usually grouped into two main traditions, called "dispositional" (or "systemic") and "aetiological". As we will argue, the autonomous perspective advocates a third one, the "organisational" view, which aims to combine the previous accounts into an integrated framework. Before expounding our own view, we shall first provide a brief overview of the other two accounts, and describe their respective strengths and weaknesses.

3.1.1 Dispositional Approaches

In the philosophical debate on functions, several authors have, against the eliminativist stance, advocated the idea that functional attributions do indeed refer to current features of the system under examination. By explicitly discarding teleology as a constitutive dimension of the concept of function, these authors hold that functions do not refer to a causal process that would explain the existence of the function bearer by appealing to its effects. Rather, functional relations are interpreted as a particular class of causal effects or dispositions of a trait – means-end relationships contributing to some distinctive capacity of the system to which they belong.[2]

The philosophical agenda of dispositional approaches focuses on providing naturalised and appropriate criteria for identifying what counts as a target capacity of a functional relationship, from which the relevant norms can be deduced, and the different dispositional approaches have proposed various criteria to identify these target capacities.

[2]On the basis of this common theoretical stance, these approaches have been labelled "causal role", "dispositional" or "forward-looking", as opposed to "backward-looking" etiological ones. Here we will use the general label "dispositional" to refer to this class of theories.

The more classical dispositional approach is the "systemic approach" (SA), which defines a function *F* as the contribution of a process *P* to a distinctive higher-level capacity *C* of the system *S* to which it belongs (Craver 2001; Cummins 1975; Davies 2001). In the SA, explaining functions means analysing a given higher-level capacity of the system in terms of the capacities of the system's components, which jointly concur in the emergence of the higher-level capacity. The SA dissolves the problem of teleology of functions by reducing them to any causal contribution to a higher-level capacity. In turn, the normative dimension of functions is reduced to the fact that the causal effect must contribute to a higher-level capacity, with no reference to a "benefit" for the system.

The explanatory strategy adopted by the SA is subject to one major criticism, namely that it seriously *under-specifies* functional ascriptions, which in turn generates several problems (see also Wouters 2005). Firstly, the SA fails to draw a principled demarcation between systems whose parts appear to have functions and systems whose parts do not (Bigelow and Pargetter 1987; Millikan 1989). Secondly, the SA lacks a principled criterion for identifying the relevant set of contributions for which functional analysis makes sense. And thirdly, the SA is unable to draw an appropriate distinction between "proper" functions and accidental, useful contributions (Millikan 1993, 2002).

Because of these fundamental weaknesses of the SA, the "goal contribution approach" (GCA) has attempted to introduce more specific constraints on what makes causal relations properly functional, by linking the concept of function to the cybernetic idea of *goal-directedness*. In particular, the GCA restricts functional attributions to causal contributions to those (higher-level) capacities that constitute the "goal states" of the system (Adams 1979; Boorse 1976, 2002; Rosenblueth et al. 1943). In particular, biological systems can be described as having the essential goal of surviving (and reproducing). Hence, biological functions are dispositions that contribute to these goals.

The main virtue of the GCA is that it provides an interpretation of functions that, in contrast to the SA, recognises and substantiates their specificity as means-end causal relationships. Nevertheless, the characterisation of a goal-directed system introduces norms whose application is not restricted to the relevant kinds of systems and capacities. As Bedau (1992) points out, the cybernetic characterisation of the goal state is unable to adequately capture the frontier between "genuinely" goal-directed systems (supposedly biological systems and artefacts) and physical equilibrium systems, which tend to some steady state or state of equilibrium (see also Nissen 1980).

Moreover, as Bedau (1992) and Melander (1997) argue, cybernetic criteria may interpret the dysfunctional behaviours of goal-directed systems as functional and, also, the GCA account lacks the theoretical resources to distinguish between functions and accidental contributions to a goal state. In sum, the GCA still seems to under-specify functional attributions, and in some cases it appears even to be a less satisfactory account than the SA.

The third main dispositional perspective proposes the identification of functions with causal contributions of components to the life chances (or fitness) of the system

(Bigelow and Pargetter 1987; Canfield 1964; Ruse 1971). Bigelow and Pargetter, in particular, have proposed the "propensity view", according to which "something has a (biological) function just when it confers a survival-enhancing propensity on the creature that possesses it" (Bigelow and Pargetter 1987: 108).

By appealing to survival in terms of enhancing propensities as the goal of a functional relation, the propensity view succeeds in restricting functions to components of biological entities. Moreover, Bigelow and Pargetter's reference to survival-enhancing *propensity* is intended to avoid functional attributions to contingent and/or accidental contributions to survival, which would be contrary to intuition and common use. Yet, as McLaughlin perceptively argues (McLaughlin 2001: 125–8), the appeal to propensities does not fully succeed in restricting functional attributions to the relevant cases. Even by restricting propensity to the current environment (the "natural habitat", in Bigelow and Pargetter's terms), it is possible to imagine, for each specific effect produced by a trait, a situation in which that specific effect would confer a (possibly low) propensity that enhances survival, and thus have a function.

The problem is that propensities to enhance survival in virtual (but not impossible) situations correspond, in a forward-looking approach, to *actual* functions of the existing trait. Moreover, to the extent that the specific contribution of the trait would presumably change in accordance with the particular condition in question, each trait in fact possesses an indefinite list of actual functions. Again, the propensity view fails to provide an adequately restricted definition of what counts as a functional relation. All (biological) functions are survival-enhancing contributions, but not all survival-enhancing contributions are functions. Appealing to propensities does not solve the problem.

To summarise, the main virtue of the dispositional approaches is their capacity to capture the fact that the concept of function points to something more than mere causal relations: functions refer to current means-end relationships, and more specifically to current contributions of components to the emergence of a target capacity of the containing system. Yet, dispositional approaches in the end fail to provide a satisfactory grounding for the normativity of functional attributions, and dispositional definitions turn out to be systematically under-specified, allowing functional ascriptions to irrelevant systems and/or capacities. In a word, the price paid for excluding the teleological dimension as a proper *explanandum* is not compensated for by a satisfactory foundation of the normative dimension.

In fact, most of the existing literature has favoured a different approach, according to which an adequate understanding of functional attributions has to deal with the problem of teleology. In particular, both the teleological and normative dimensions are conceived as being inherently related to the *aetiology* of the functional trait.

3.1.2 Aetiological Theories

The mainstream philosophical theory of functions is the aetiological approach (Wright 1973, 1976; Millikan 1984, 1989; Neander 1980, 1991; Godfrey-Smith

1994). The aetiological approach defines a trait's function in terms of its aetiology (i.e. its causal history): the functions of a trait are past effects of that trait that causally explain its current presence. In sharp contrast with dispositional accounts, the aetiological approach explicitly takes the issue of teleology as the central problem of a theory of functions.

The first aetiological approach was proposed by Wright, who defined functions as follows:

The function X is Z means:

1. X is there because it does Z.
2. Z is a consequence (or result) of X's being there (Wright 1976: 48).

Wright's definition explicitly appeals to a form of causal loop, in which the effect of a trait helps to explain – teleologically – its existence. The scientific validity of Wright's definition has been questioned and, moreover, several obvious counterexamples have been formulated (see, for instance, Boorse 1976).

In order to ground the teleological dimension of functions without adopting an unacceptable interpretation of the causal loop described by Wright, mainstream aetiological accounts, usually called "selected effect (SE) theories", have appealed to the Darwinian concept of Natural Selection as the causal process, which would adequately explain the existence (or, more precisely, the maintenance over time) of the function bearer by referring to its effects. The gist of SE theories is that functional processes are not produced by the same tokens whose existence they are supposed to explain. Instead, the function of a trait is to produce the effects for which past occurrences of that trait were selected by Natural Selection (Godfrey-Smith 1994; Millikan 1989; Neander 1991). Selection explains the existence of the *current* functional trait because the effect of the activity of *previous* occurrences of the trait gave the bearer a selective advantage. The main consequence of this explanatory line is its historical stance: what makes a process functional is not the fact that it contributes in some way to a present capacity of the system, but rather that it has the right sort of selective history.

By interpreting functions as selected effects, SE theories are able not only to deal with the problem of teleology, but also to ground the normativity of functions. By defining functions as effects subject to an evolutionary causal loop, SE theories identify the norms of functions with their *evolutionary conditions of existence*: the function of a trait is to produce a given effect because *otherwise*, the trait would not have been selected, and would not therefore exist.

Several virtues of SE theories are often emphasised, including their capacity to exclude functional attributions to traits of physical systems, and their ability to unambiguously identify functions from among the whole set of all processes occurring in a system and to draw a boundary between functions and accidental useful effects. Nevertheless, SE theories have their own weaknesses, which have been extensively discussed in the literature (see, for instance, Boorse 1976; Cummins 2002; Davies 1994, 2000). We will focus here on one specific weakness of the theories, which Christensen and Bickhard (2002) have labelled their *epiphenomenalism*. The crucial drawback of SE theories' explanatory line is the implication that functional attributions bear no relation to the *current* contribution of the trait to the system, since they point solely to the selective history of the trait. This is at odds with the fact that functional attributions to biological structures do seem to bear some relation to what they currently do, and not only to what explains their current existence.

To solve some difficulties inherent to previous formulations of aetiological theories (mainly that they attribute proper functions to effects that are, in fact, no longer functional in the current system), Godfrey-Smith (1994) has proposed a "modern history theory" of function. In his approach, functions are "dispositions or effects a trait has which explain the recent maintenance of the trait under natural selection" (Godfrey-Smith 1994: 199; See also Griffiths 1993). While it successfully counters several objections raised against previous versions of the theory, Godfrey-Smith's account is no better placed to deal with the problem of epiphenomenalism. More precisely, as McLaughlin (2001: 116) points out, by reducing the cases in which it attributes functions to currently non-functional traits, Godfrey-Smith's account (which is explicitly an historical one) possibly reduces "uncooperative cases", but does not provide a principled solution to the problem.

Accordingly, SE theories provide an account that is problematically epiphenomenal, in the sense that it maintains that the attribution of a function does not provide information about the current system being observed. From the perspective of SE theories, a function does not tell us anything about the current organisation of the system being analysed.

3.2 The Organisational Account of Functions

The outcome of this brief critical survey is that current theories of functions seem to face a dilemma, arising from the way in which they deal with the two main issues related to the concept of function, i.e. its teleology and its normativity. Dispositional theories try to account for functions in terms of current contributions to some target capacity of a system, and discard the teleological dimension, but seem unable to provide fully adequate normative criteria for functional attributions. Aetiological theories, on the other hand, try to account for both the teleological and normative dimensions of functions, but appear inevitably historical and are unable to justify how functional attributions may refer to features and properties of the current system.

According to some authors, the solution to the dilemma consists of concluding that there is no unified account of functions, and that aetiological and dispositional approaches provide two different yet complementary concepts of function (Allen and Bekoff 1995; Godfrey-Smith 1994; Millikan 1989). Other authors, such as Kitcher (1993), Walsh (1996), and Walsh and Ariew (1996), have claimed that there is, in fact, a single concept of function, in which the aetiological and dispositional formulations can be subsumed as special cases. In this section, we argue that, from the autonomous perspective, there is indeed room for a unified account of functionality, based on the properties of self-determination of biological organisation.

The core of the organisational account (OA) is the idea that functional ascriptions do account *at the same time* for both the existence of functional traits and their current contribution to a system capacity, since functions make sense only in relation to the specific kind of organisation which is characteristically at work in biological organisms. In particular, as we shall argue, functions correspond to those causal effects exerted by the constraints subject to closure that contribute to maintaining the organisation.

Before expounding our own version of the OA, it should be mentioned that, very recently, a considerable amount of work has been done in this direction by Bickhard (2000, 2004), Schlosser (1998), Collier (2000), McLaughlin (2001), Christensen and Bickhard (2002), Delancey (2006), Edin (2008), and more recently by ourselves (Mossio et al. 2009b; Saborido et al. 2011; Saborido and Moreno 2015). In spite of some differences between the various formulations, there seems to be substantial convergence[3] regarding the fundamental tenets of the OA, which makes it a credible philosophical alternative to both aetiological (mainly in its "selected-effects" version) and systemic-dispositional accounts.

3.2.1 Teleology, Normativity and Self-Determination

The OA relies on an understanding of biological systems as sophisticated and highly complex examples of natural self-maintaining systems. In particular, the first claim of the OA is that self-determination, as characterised in Chaps. 1 and 2, constitutes the relevant emergent causal regime in which the teleological and normative dimensions of functions can be adequately naturalised.

On the one hand, the causal regime of a self-maintaining system provides a naturalised grounding for the teleological dimension. Since the activity of the system S contributes, by exerting a constraint on its surroundings, to the maintenance of some of the conditions required for its own existence, the question "Why does S exist?" can be legitimately answered by "Because it does Y". This justifies explaining the existence (again, in the specific sense of its *maintenance* over time) of a system in "teleological" terms by referring to its causal effects.

On the other hand, self-maintenance grounds normativity. The activity of a self-maintaining system has an intrinsic relevance for itself, to the extent that its very existence depends on the constraints exerted through its own activity. Such intrinsic

[3]Christensen and Bickhard (2002) have suggested, relying on their own work on the notion of biological autonomy, that the organisation of autonomous systems provides an adequate grounding for the normativity of functional attributions. In a similar vein, McLaughlin (2001) has developed an account in which both the teleology and normativity of functions can be naturalised in the organisation of self-reproducing systems. Despite some terminological differences, the central idea of these approaches (i.e. that the organisational closure instantiated by living systems provides an adequate basis for naturalising functions) fundamentally coincides with that defended here, and we explicitly recognise this theoretical relationship.

relevance generates a naturalised criterion for determining what norms the system is supposed to follow: the system must behave in a specific way, otherwise it would cease to exist. Accordingly, the activity of the system becomes its own norm or, more precisely, its conditions of existence are the intrinsic and naturalised norms of its own activity.

Note that, so far, we have been generally referring to self-maintenance, and not closure. Hence, we acknowledge that the grounding of the teleological and normative dimensions goes beyond the biological domain, and includes some kinds of physical and chemical self-maintaining systems. Let us take the simple example of a candle flame. As Bickhard (2000: 114) points out, by constraining its own surroundings, the flame makes several contributions to the maintenance of the conditions required for its own existence. Indeed, the flame keeps the temperature above the combustion threshold, vaporises wax and induces convection (which pulls in oxygen and removes combustion products). Accordingly, to the question "Why does the flame exist?" it is legitimate to answer "Because it does X": the existence of the combustion reactions (the flame itself) is explained (at least in part) by taking into account the effects of its constraining action. Moreover, what the flame does is relevant and makes a difference for itself, since its very existence depends on the specific effects of its activity. The conditions of existence of the flame are the norms of its own activity: the flame must behave in a specific way, otherwise it would disappear.

One may object that, if self-maintenance as such provides the relevant grounding for teleology and normativity, then the OA should allow functions to be ascribed to physical dissipative systems. But of course, this implication seems unsatisfactory since, usually, no one ascribes functions to physical systems. Hence – the objection could continue – the OA clearly fails to restrict functions to the relevant kind of systems, just as dispositional approaches do. To this objection, we reply by formulating the second claim of the OA, according to which self-maintenance is a necessary but not sufficient condition for grounding functions in a naturalised way. Functions emerge when the self-maintenance is realised in the specific form of closure.

3.2.2 *Closure, Organisation and Functions*

The second claim of the OA is that when self-maintenance is realised as closure, then the causal effects of the constraints subject to closure are functional. Accordingly, as we claimed in Chap. 2, functionality is an emergent property of closure. Closure of constraints is therefore closure of functions.

Before providing a more precise definition and exploring some implications, let us clarify what is behind the conception of functionality advocated by the autonomous perspective.

The central idea is that functionality, in addition to teleology and normativity, includes a third dimension, that of *organisation*. Functions, we submit, involve the fact that self-determination is achieved through the interplay of a network of mutually dependent entities, each of them making *different* yet *complementary* (and also *hierarchical*, as in the cases of regulation and control, discussed in Chap. 1, Sect. 1.8) contributions to the maintenance of the boundary conditions under which the whole system can exist. In other words, to ascribe functions we must distinguish between different causal roles in the system, a division of labour among the parts. And, of course, this is precisely what happens when closure of constraints is realised. As clarified above, closure is realised as the mutual dependence of the whole set of constraints which collectively achieve self-determination. But the very idea of mutual dependence presupposes that the various constraints produce different yet complementary causal effects: if all constraints produced the same effect, they would not depend on each other, and each constraint would be able to self-maintain individually. That is why, in our view, functions are not ascribed to dissipative structures. As discussed earlier, in this case there is only a single entity (the macroscopic structure itself) that acts as a constraint on the surroundings, and contributes to maintaining the conditions of its own existence. Since there is no need to distinguish between different contributions to self-determination generated by different constraints, functional ascriptions are meaningless.

At this point, it is important to set out one general implication of the autonomous perspective. The concepts "closure" and "organisation" are inherently linked. In the technical sense defined in Chap. 1, an organisation appears precisely when a set of constraints realise closure. Here, we add a third dimension. To the extent that closure is taken as the naturalised ground of functions, it follows that the concept of functionality itself is theoretically linked to that of closure and organisation. "Functionality", "closure", and "organisation" are then *mutually related concepts*, which refer to the very same causal regime; in other words, in the autonomous perspective an organisation is by definition closed and functional.[4]

Functional ascriptions and explanations are relevant as soon as the kind of organisational complexity realised by closure comes into being. Accordingly, it might be useful to focus on the distinction between the organisational and what could be labelled the "material" complexity of a system, i.e. the variety of its internal components. Minimal self-maintaining systems may indeed differ considerably with respect to their material complexity. Whereas many physical dissipative systems possess a rather homogeneous nature in terms of the variety of molecules of which they are made up (e.g. whirlwinds and Bénard cells), other systems, including chemical dissipative systems such as candle flames, have many different molecular components. Certain types of dissipative chemical systems (the Belousov-Zhabotinsky reaction, for instance) may even possess a high degree of material complexity.

Even high material complexity, however, has nothing to do with organisational closure, and therefore does not imply functions. In the case of the flame, for instance, the different chemical components all "converge" to generate a single macroscopic

[4]Of course, the reciprocal equivalences hold equally: closure refers to a functional organisation, and functionality indicates a closed organisation.

pattern (the flame), which in turn constrains the surrounding dynamics. Accordingly, it is not possible in this case to distinguish between the different ways in which the various components contribute to the self-maintenance of the system. The flame, although materially quite complex, is organisationally simple: in fact it has no organisation at all. Hence, functional attributions to components of the flame, as well as to all physico-chemical dissipative structures, are not meaningful. The realisation of closure requires not only that different material components be recruited and constrained to differentially contribute to self-maintenance but, in addition, that the constraints which contribute to self-determination be generated, and maintained, within and by the organisation of the system.

Let us now give an explicit and formal definition of function. According to the organisational account, a trait T has a function if, and only if, it exerts a constraint subject to closure in an organisation O of a given system. This definition implies the fulfilment of three different conditions (Saborido et al. 2011):

C_1. T exerts a constraint that contributes to the maintenance of the organisation O;
C_2. T is maintained under some constraints of O;
C_3. O realises closure.

Let us apply this definition to the classic example of the heart. The heart has the function of pumping blood since (C_1) pumping blood contributes to the maintenance of the organism by allowing blood to circulate, which in turn enables the transport of nutrients to and waste away from cells, the stabilisation of body temperature and pH, and so on. At the same time, (C_2) the heart is maintained under various constraints exerted by the organism, whose overall integrity is required for the ongoing existence of the heart itself. Lastly (C_3), the organism realises closure, since it is constituted by a set of mutually dependent structures acting as constraints, which, by contributing in different ways to the maintenance of the organisation, collectively realise self-maintenance.

It should be underscored that this characterisation of functions is consistent with the one proposed by Wright. In this example, the heart is there because it pumps blood (otherwise the organism, and thus the heart, would disappear), and pumping blood is a consequence of the heart's being there. This consistency stems from the fact that the organisational account, by appealing to a causal loop at work in the organisation of the system, provides an argument for naturalising both the teleology and normativity of functions, which, at an organisational level, mirrors the explanatory strategy adopted by the aetiological approaches. The resulting account represents an integration of the aetiological and dispositional perspectives, since it may at the same time explain the existence of the trait and its current contribution to the maintenance of the system.[5]

[5]In a recent contribution, Artiga (2011) offered a detailed critical analysis of the organisational account. Some of his remarks have been taken into account in the present formulation of the OA, while others (with which we do not agree) would require a full reply; but we will leave this for a future analysis.

The organisational definition given above is very general, and aims at encompassing all particular cases. Yet, actual functional ascriptions would take into account the complexity of autonomous organisation. This means, first of all, that functional ascription could vary according to the specific instance of closure that the system is realising at a given moment (what we called a "regime of self-maintenance"). Also, for each specific constitutive regime, as discussed in Chap. 1, Sect. 1.8 above, autonomous systems can realise different *orders* of closure, in particular insofar as regulation is involved. Moreover, in Chap. 6 we will discuss how different *levels* of closure (and then of organisation) can be described in certain classes of biological organisms, in particular multicellular ones.

Each specific regime, orders and levels of closure generate, as argued in this chapter, a distinct set of norms and functions. For instance, a given function could be related either to an individual cell (first-level) or to the whole multicellular organism (second-level) to which that cell belongs; in each of these cases, that very function could be either constitutive (first-order) or regulative (higher-order). And that function could be at work only within a specific regime of maintenance of the considered system, realised, for instance, only in some particular conditions or at a given moment. As a consequence, adequate functional ascriptions should make explicit, in each specific case, which are the regime, order, and level of the closure involved in C_3.

Lastly, it should be noted that, in principle, for each constraint subject to closure, functional ascriptions may concern either the *structure* itself (the trait) or the *effects* produced by that structure. Although the second option would possibly be more precise, here we mainly refer to the functions of traits and structures, which is consistent with the typical use of functional ascriptions in the relevant literature, as well as in ordinary language (see also Wimsatt 2002: 179).

3.3 Implications

The organisational account of functions has several relevant implications for the philosophical debate. Some of them[6] have already been spelled out in a previous study (Mossio et al. 2009b), and shall not be discussed here. In this section, we will focus on two main issues that are of crucial importance for assessing the scope and prospects of the OA: the ascription of cross-generation functions and the characterisation of malfunctions.

[6]For example, the distinction between functionality and usefulness; or the relationship between the concept of primary functions and the aetiological concept of proper functions.

3.3.1 *Cross-Generation Functions*

A major theoretical challenge facing the organisational account concerns, as Delancey (2006) has argued, the capacity to ground "cross-individual functions", i.e. those functions which go beyond the boundaries of individual biological systems. Let us explain what exactly this challenge consists of.

In the OA, functions are characterised as contributions of parts to closed organisations, and since closed organisations are typically realised by individual organisms, the OA appears to have trouble grounding those functions involving several individuals and their interactions. In particular, it is unclear whether and how the organisational approach would account for what Schlosser (1998) calls "cross-generation functions", for instance, the function of reproductive traits (e.g. the function of semen to inseminate the ovum). In these cases, in fact, the trait seems to contribute to maintaining the organisation of a system that is different from the system of which it is a component. Hence, the trait does not contribute either to the maintenance of an organisation or to its own self-maintenance. Still, we do ascribe cross-generation functions, just as we do, for instance, in the case of the reproductive function of semen. At first sight, then, cross-generation functions constitute a major group of counterexamples within the organisational approach.

As we explained in a previous work (Saborido et al. 2011), some of the authors who advocated the organisational account were of course aware of this issue, and proposed (following very different paths) solutions, which were designed to enable the account to embrace both intra- and cross-generation biological functions. Broadly speaking, the existing formulations can be regrouped into two main versions. The first version, advocated by Schlosser (1998) and McLaughlin (2001), tends to characterise reproductive functions as states or processes, which are causally required for the reproduction of the trait that causes them. The emphasis is therefore on the self-reproduction of the trait, rather than specifically on the whole system that, nevertheless, must possess the adequate properties to enable trait self-re-production. The second version, proposed by Collier (2000), Christensen and Bickhard (2002), shifts the focus onto the organisation of the system, and interprets reproductive functions as contributions to a higher-level self-maintaining organisation.

Delancey's analysis criticises all these "unified accounts" by pointing out their weaknesses and drawbacks. As an alternative, he proposes a "splitting account", according to which intra- and cross-generation functions are in fact two different kinds of biological functions, requiring different conceptual treatment within an organisational account.[7]

[7]We do not describe Delancey's account here. For details, see Saborido et al. 2011.

It is our contention, however, that the OA may provide a unified definition applying to both intra- and cross-generation functions. The essence of our argument will be that cross-generation functions contribute to the maintenance of systems, which realise a closed self-maintaining organisation in the very same sense as that of systems whose parts are ascribed intra-generational functions. To the extent that the two kinds of systems do not differ in terms of organisational self-maintenance, there is no need to invoke two kinds of functions, and the ontological problem is therefore overcome.

Before developing our own formulation, let us briefly discuss another proposal, put forth by Christensen and Bickhard (2002), which also tries to provide a unified account of intra- and cross-generation functions within the autonomous perspective. Their central move consists of appealing to higher-level organised self-maintaining systems, composed of individual organised self-maintaining organisms, in which reproductive traits could be subject to closure. In particular, Christensen and Bickhard explicitly grant systems like populations or species the status of autonomous[8] systems, making them relevant supports for functional ascriptions, just like individual organisms:

> Living organisms in general are autonomous systems, as are reproductive lineages, species, and some kinds of biological communities (Christensen and Bickhard 2002: 3).

As a consequence, intra- and cross-generation functions are simply contributions to the maintenance of different specific systems, sharing the same kind of organisation at different scales. Whereas intra-generation functions would contribute to the autonomous organisation of individual organisms, cross-generation functions would contribute to the autonomous organisation of the lineage, the species or the biological community in question.

Christensen and Bickhard offer an elegant alternative to the splitting account by admitting the idea of higher-level autonomous systems, namely, systems that would include individual organisms as parts, and that would ground the ascription of cross-generation functions. Accordingly, the heart is functional because it contributes to the autonomy of each individual vertebrate organism, while semen is functional because it contributes to the autonomy of the species.

Yet, this solution is problematic, as Delancey's lucid criticism (Delancey 2006) shows. As he points out, considering those higher-level systems that are relevant for grounding cross-generation functions as autonomous systems does not come without a price. Whereas an individual organism is a paradigmatic case of an autonomous system, "the sense in which the species or some population is a complex system of the appropriate kind is much more difficult to discern" (Delancey 2006: 90). For instance (and the list could be longer), such higher-level systems

---

[8]Christensen and Bickhard use the term "autonomy" in a somewhat weaker sense than the one developed in this book. Note that, in our account, closure is a *sufficient* requirement for grounding functions: in other words, functional systems are not necessarily autonomous systems.

have no clear boundaries, no stable form and, above all, it is very hard to see how their own "internal" organisation would realise closure, as is the case for individual autonomous systems.

According to Delancey, the organisational account has not explored these radical differences with sufficient accuracy, which means that the interpretation of higher-level systems as autonomous (or at least closed) systems appears, to say the least, to be an ad hoc hypothesis to cover reluctant cases.[9] In particular, to the extent that Christensen and Bickhard appeal to the idea of autonomy in a fairly general sense, we assume that Delancey's criticism applies equally to an interpretation of higher-level systems as organised self-maintaining systems, which could be put forward within our own conceptual framework.

A possible reply would consist of arguing that other biological supra-organismal systems do possess the properties required to be considered self-maintaining organisations. Let us briefly explore another possibility, not mentioned by Delancey's analysis: the ecosystem. Compared with species, lineages or populations, there do indeed seem to be better reasons for considering ecosystems higher-level closed systems, relevant for functional ascriptions, especially if one adopts our characterisation in terms of self-maintaining organisations realising closure, rather than the more demanding terms of autonomy. Although there are clear differences (just to mention one: the ecosystem has no physical boundaries), ecosystems share several organisational properties with individual organisms. For instance, the various components (be they individual organisms or groups of organisms) contribute to maintaining a global organisation (the ecosystem itself), which in turn is a general condition for their own continuous existence. Similarly, the various components seem to be mutually dependent, so that the disappearance, death, or anomalous behaviour of one may provoke the collapse of the whole ecosystem.

For these and other reasons, the ecosystem has some features in common with an organism, and in fact it does not seem unreasonable, despite being somewhat uncommon, to use a functional discourse to describe it. So, for instance, we could describe and explain the organisation of an ecosystem by attributing to its various components functions such as the regulation of air, climate, water, water supply, disturbance prevention, soil formation and erosion, nutrient cycling, waste treatment, pollination, biological control of pests and diseases, and so on (De Groot et al. 2002; Nunes et al. 2014). Specifically, cross-generation traits would have the function of regenerating the various components of the ecosystem, which would tend to decay because of their dissipative nature.

In our view, the idea that the ecosystem is, at least, a closed self-maintaining system is an attractive one, and deserves further investigation. Indeed, we discuss

[9]Delancey's remark is fundamentally correct. As a matter of fact, we try to make some preliminary steps towards an account of higher-level closed organisations at the end of Chap. 4, and then of higher-level autonomous systems in Chap. 6.

this issue in more detail in Chap. 4, Sect. 4.5.[10] Yet, the search for higher-level closed organisations would be largely irrelevant for solving the problem of cross-generation functions within the organisational account since, we submit, *the reason why we ascribe functions to cross-generation traits is not related to their contribution to the maintenance of some higher-level system.* Cross-generation functions, we argue, do not require an account of higher-level closed systems in order to be adequately naturalised within an organisational account. Let us then turn to our proposal.

The gist of our account consists of arguing that the apparent difficulty in integrating cross-generation functions into the definition does not stem from an ontological difference between intra- and cross-generation functions but rather from an inadequate understanding of what a closed self-maintaining organisation actually is. Cross-generation functions constitute a "recalcitrant" class of functions only if the boundaries of the self-maintaining organisation are confused with the boundaries of the individual organisms themselves, whereas, in fact, they are conceptually distinguishable. Once this confusion has been cleared up, the ontological problem disappears.

In our account, functional traits are those traits that, by being subject to closure, contribute to the maintenance of an organisation, which in turn exerts some causal influence on the production and maintenance of the traits. The whole system, as discussed in Chap. 1, realises a self-maintaining organisation through closure. The first remark is that a self-maintaining organisation occurs in time, and can be observed only in time. Now, as we have mentioned in Chap. 1, Sect. 1.6, biological organisms undergo various material, structural and morphological changes and modifications over time. If, due to these changes, one were to consider the various temporal instances O_1, O_2, . . .O_n, as different organisations, then functions could not exist. A trait would be produced by a given organisation O_1, and would contribute to maintaining another organisation O_2. No organisation would actually self-maintain, no trait could be subject to closure, and functions could not be ascribed.

The crucial point is that, in the organisational account, these changes are irrelevant with regards to functional ascriptions, because what matters is the *continuity of organisational closure*.

[10]Besides, the claim that certain supra-organismal organisations could harbour functional relations does not undermine our previous proposal of grounding functions in the causal regime of organisms. Since obviously any supra-organismal organisation requires the existence of organisms, it implicitly supposes the (intra)organismic organisation in order to ground the existence of functions. For example, the constraints that ensure the maintenance of an ecosystem are generated by the specific metabolic organisations of different types of species in a given ecosystem. In this sense, the requirement that the constraints be generated within the system – if by the system we understand the supra-organismal organisation– is only satisfied partially (Nunes et al. 2014).

As discussed in Chap. 1, Sect. 1.6, the realisation of closure requires considering a *minimal* temporal interval (say, τ_n), wide enough to include the specific time scales[11] at which all constitutive constraints and their mutual dependencies can be described. As a consequence, the various temporal instances (at time scales τ_1, $\tau_2 \ldots < \tau_n$) of a system can be considered – in spite of any changes that may occur – instances of the *same* encompassing self-maintaining organisation, to the extent that their constitutive organisation realises closure at τ_n. In particular, this implies that the system in which a trait x performs an enabling function at time τ_1 is the *same* system in which, at τ_2, that function of x is dependent, if both τ_1 and τ_2 are included in τ_n (at which closure is realised) .[12]

In other terms, for the purposes of ascribing functions, the continuity of closure (and thus the maintenance of the system) takes precedence as a criterion of individuation over other criteria on the basis of which the various instances of the organisation would possibly *not* be taken as instances of the same system. If there is a causal dependence between two temporal instances of a system, such that their conjunction realises closure, then it could be claimed that, in this respect (and possibly *only* in this respect) the two instances are temporal instances of the same encompassing organisation.

Our central thesis is that self-maintaining organisations, which ground the ascription of cross-generation functions, and specifically reproductive functions, comply with the very same characterisation as those organisations, which ground intra-generation functions. While they may (and actually do) differ in important ways, the two classes of self-maintaining organisations do not differ with respect to the relevant properties that ground functional ascriptions.

Cross-generation functions are subject to closure within those self-maintaining organisations whose extension in time goes beyond the lifespan of individual organisms. For instance, by inseminating the ovum, mammalian semen contributes to the maintenance of the organisation by contributing to the production of a new individual organism to replace the previous one. In turn, the organisation (which consists in the conjunction of both the reproducer and the reproduced system) exerts several constraints under which the semen is produced and maintained. The crucial point is that the organisation of the system constituted by the conjunction of the reproducing and reproduced organisms (in this specific case, a minimal lineage with two elements) has exactly the same status, in terms of self-maintenance, as that of the individual organisms themselves. The fact of considering the organisation

[11]Of course, time scales may greatly vary according to the specific function: the function of the lung is subject to closure in a very short period of time (one cannot stop breathing for more than a few minutes) whereas, for instance, the function of the stomach is subject to closure over a longer period of time (one can stop eating for days).

[12]See Chap. 1, Sect. 1.5, for the definition of dependence among constraints (functions), as well as the distinction between enabling and dependent.

of individual organisms (at τ_n) or their conjunction (at τ_{2n}) as the relevant self-maintaining organisation depends on the explanatory exigencies for functional ascriptions.

Since what matters in the case of organisational self-maintaining systems is the fact that they use their own constitutive organisation to exert a causal influence on the maintenance of (at least part of) their own conditions of existence, then the organisation of the "encompassing system" made up by a reproducer and a reproduced system itself fits the characterisation of a closed self-maintaining organisation. Reproduction, in this sense, simply constitutes one of the functions through which the organisation succeeds in maintaining itself beyond the lifespan of individual organisms. Since the encompassing system made up by the reproducer and the reproduced organism possesses a temporally wider self-maintaining organisation, reproductive traits are subject to organisational closure, and their functions are correctly grounded in the organisational account.

Why do cross-generation functions appear problematic? Intuitively, the point seems to be that reproduction involves a dramatic transition from the reproducer to the reproduced organism, so much so, in fact, that it cannot be maintained that they constitute the same organised system. Given that reproduction may involve phenomena like embryogenesis and development, such causal and phenomenological discontinuities prevent us from considering these systems as temporal instances of the same self-maintaining system. Only individual organisms are genuine self-maintaining organised systems.

In our view, this objection is not compelling, since it is based on an insufficient understanding of what matters for considering that an organisation is self-maintaining. The crucial requirement, as discussed above, is the functional dependence of the temporal instances of an organisation. Two systems which realise closure at a time scale τ_n, may be said to constitute, at a longer time scale τ_{2n}, two temporal instances of an encompassing self-maintaining organisation if it can be shown that the conjunction of the two instances realises itself closure (which includes more functions, in particular cross-generation functions). The relevant question is: is there a causal dependence between the two instances, such that the encompassing organisation can be said to realise closure? Or, to put it another way, is there continuity in the realisation of closure across the successive instances of the self-maintaining organisation? Since the answer to these questions is, in a fundamental sense, affirmative for the case of the relationship between the reproducer and the reproduced system, we claim that the encompassing organisation including them is itself a closed self-maintaining organisation that maintains itself also through reproduction.

As Griesemer has pointed out, the reproduction process does indeed involve the material connection between the reproducer and the reproduced system:

> Reproduction, ... is the multiplication of entities with a material overlap of parts between parents and offspring. Material overlap means that parts of the parents (at some time) become parts of the offspring (at some other time). Thus reproduction is no mere transmission or copying of form– it is a flow of matter (Griesemer 2002: 105).

Rather than a flow of matter as such, the autonomous perspective emphasises the continuity of the functional organisation, which maintains itself over time, also because of reproduction. As it has been argued (Zepik et al. 2001) the occurrence of reproduction may be explained in terms of the time relation between the production and decay of the constitutive components in a far-from-equilibrium organisation. If the rate of replacement of the constitutive components is faster than their decay, the self-maintaining cycles of the system will prompt it to establish reproductive cycles: the system will grow and reproduce; otherwise, it will disintegrate. Only in the very unlikely case of coincidence between the rates of replacement and decay will the self-production cycles of the system realise self-maintenance without reproduction.

The macroscopic transition produced by the reproductive process can then be seen as the way in which the organisation actually manages to self-maintain beyond the temporal boundaries of individual organisms. Just as the various temporal instances of an individual organism are considered, despite changes and modifications, a single self-maintaining organisation to the extent that the organisational properties are causally linked throughout the various instances, so too are the various instances of the inter-generational organisation considered a single self-maintaining organisation due to causal dependence between the instances. The conceptual operation is exactly the same, the difference lies only in the level of temporal "zoom" through which self-maintenance is observed.[13]

This is why development is an essential feature of the self-maintaining organisation of living organisms. Once we see reproduction as a process that causally connects the reproducer and the reproduced organisations, development appears as a necessary step in this continuous process of complex self-maintenance. Indeed, self-maintenance of biological individuals can only be ensured through a continuous unfolding of changes, including reproduction and development. Chapter. 6 will further elaborate on the place of development within the theory of autonomy.

Since the only relevant ground for functional ascriptions is organisational closure, all other criteria of distinction between biological systems may be considered as irrelevant for this specific purpose. This is why reproductive traits can be said to be subject to organisational closure and why, then, we ascribe functions to them.

3.3.2 *Malfunctions*

A second major implication of the organisational account is the characterisation of malfunctions. It is often claimed (see for instance Neander 1995; McLaughlin 2009; Krohs 2010, 2011; Christensen 2012) that a satisfactory theory of functions

---

[13]The fact that self-maintenance, in the form of closure, spans beyond the lifetime of individual organisms is an important aspect related to the historical dimension of autonomy. See Chap. 5 for a detailed discussion.

should be able to ground both functions and malfunctions,[14] since a function can be performed well, or defectively, or even not at all. Yet, in spite of the fact that the concept of malfunction is widely used both in everyday language and in scientific disciplines such as physiology or medicine,[15] the theoretical grounding of malfunctions has received little attention in the philosophical debate, which has mainly focused on the concept of function.

What is the gist of a philosophical account of malfunction? Claiming that a trait can function "well" or "poorly" implies a reference to a norm, which may or may not be fulfilled. Malfunctions, then, have a normative dimension, just as functions do. But, and here comes the central philosophical issue, the norms grounding functions and malfunctions are not the same, and an independent justification must be provided for each.

The closure of biological organisation provides the relevant grounding in which the concept of function can be adequately naturalised. In particular, it generates the norms that the traits subject to closure must fulfil in order to be functional: as we claimed, the organisational approach identifies these norms as the conditions under which the whole organisation (or, more precisely, each specific regime of organised self-maintenance, see Chap. 1, Sect. 1.8.1), and consequently each of its constituents, can exist. Thus, functional traits are all those whose causal effects contribute to the maintenance of the whole organisation.

Now, of the whole set of traits that fulfil the norms of functionality, some do so well and others poorly. Yet the norms generated by closure are blind with respect to the distinction between these two types of effects, because both of them contribute to the maintenance of the organisation (albeit in some cases poorly), and both are therefore *functional*. Hence, the distinction between (well-)functions and malfunctions requires an additional set of norms, on the basis of which it might be possible to discriminate between different ways of contributing to the maintenance of a closed organisation.

One important implication of this line of thought is that functions and malfunctions are by no means alternative kinds of entities; rather, malfunctions are a subset of functions that, while fulfilling the norms generated by closure, fail to comply with the norms of *well*-functions. This enables, among other things, a straightforward conceptual distinction to be made between *malfunctions* and *nonfunctions* (often confused both in ordinary use and specialised literature): while the former are indeed a class of functions, the latter do not. Nonfunctions refer to the effects of traits which do not comply with the norms generated by closure, and do not therefore contribute at all to maintaining the organisation. A kidney that does not filter blood, for instance, is nonfunctional rather than malfunctional. The distinction between

[14] We prefer to use the term *malfunction*, because *dysfunction* is usually used to refer both to malfunctional and nonfunctional behaviours.

[15] The concept of malfunction has often been used to justify the conceptual distinction between health and disease: some of the most influential groundings of the concept of disease have specifically interpreted diseases as malfunctions (Boorse 1977, 2002; Schramme 2007).

nonfunctions and malfunctions also serves to highlight the fact that malfunctionality is a *matter of degree* (Krohs 2010: 342). While functions are all-or-nothing concepts (a trait is either functional or nonfunctional), malfunctions admit degrees and a given trait can contribute more or less well (or poorly) to the maintenance of the organisation.

How does the organisational account deal with the concept of malfunction? Although no fully-fledged organisational definition of malfunction has been proposed so far, several authors whose approach could be considered within, or at least close to, the organisational account have pointed to a link between malfunction and adaptivity. For instance, Edin (2008) refers to malfunctions in terms of deviations from the "optimal self-maintenance" of a living system:

> Organisms are typically endowed with multiple, extensive and complex feedback systems, many of which have a set point that, when considered from the standpoint of the maintenance of the organism, is close to optimal. For this reason, physiologists talk about events or circumstances that cause the variable magnitude of such a system to deviate from the set point as disturbances or challenges. (Edin 2008: 206)

Christensen and Bickhard (2002) also consider malfunctionality to be related to the adaptive properties of organisms:

> There are a number of reasons why understanding the relative significance of dysfunction is an important adaptive issue. It is important to understand the wider systemic implications of failure in order to understand whether and how the system can compensate. It is also important to know how the system can recognise failure as part of its compensatory abilities. These are surely important issues for understanding functional organisation (Christensen and Bickhard 2002: 18).

In what follows, we will elaborate on this very idea, by relying on our characterisation of regulation exposed in Chap. 1, Sect. 1.8.2 above. As we discussed, biological organisms have to modulate their organisation to cope with the changes that they continuously undergo, be they internally or externally generated (for instance, in this second case, by a variation of the environmental conditions). Regulation is a specific form of modulation, such that a functional subsystem (a dedicated mechanism) of the organisation induces the establishment of a different and more adequate constitutive regime of self-maintenance, among a set of possible ones. Regulatory functions are, then, second-order functions (subject to second-order closure and norms) that modulate the constitutive set of functional traits and their interrelations.

In a nutshell, our account of malfunction is the following. The whole dynamic repertoire of the constitutive organisation on which regulation is exerted is limited by its physical and material structure, which implies, in particular, that each trait can only operate within a given potential range of activity. For each specific regime of self-maintenance that the system may adopt, a specific *admissible* range of activity, included in the potential one, can be determined.

If, because of some structural defect, a particular trait (1) does not modulate its activity in spite of the triggering of a regulatory mechanism and (2) as consequence, it is unable to operate within the admissible range determined by some of the regimes of self-maintenance among which regulation governs the shifts, then the trait malfunctions in organisational terms. Let us explain this idea in more detail.

Within each specific realisation of a closed organisation (i.e. each regime of self-maintenance), functional traits *presuppose*[16] each other, which means that the whole set of mutual interactions among them determines the range of admissible functional effects, defined as a subset of all potential effects that the trait may possibly produce, given its own structure. For example, a human heart can pump blood within a certain range of potential frequencies, among which a range of admissible frequencies are determined by each ongoing realisation of the organisation. Similar ranges apply of course to the lungs, kidneys . . . and to all other organs and functional traits.

Suppose that, in some circumstances, a regulatory mechanism is triggered to shift an organism form a given regime of self-maintenance to a different one. For instance, the autonomic nervous system (the regulatory subsystem, in this case), in a situation of danger, can send signals to move from a regime "at rest" to another one "under stress" in which the organism runs. Suppose also that, for some structural reason, one functional part of the organism does not modulate its activity and, as a consequence, it is unable to match the functional presuppositions of the regime induced by the regulatory functions. For instance, the coronary artery might not be able to increase its diameter sufficiently to match the higher rate of blood flow pumped by the heart: as a consequence, its range of activity is not in accordance with the functional presuppositions of the other functional traits and organs in this specific circumstances. Regulatory functions might therefore fail in modulating the defective trait's activity, so to match the new functional presuppositions (fall within the admissible ranges). For that specific trait, in a word, regulation had no effect.

In these specific situations, in which an unresponsive trait does not modulate its activity as required by the intervention of regulatory functions and therefore prevents adaptive regulation to shift to a different first-order organisational regime, so that the whole system can only remain in a specific organisational regime in which the trait match the functional presuppositions, that trait is malfunctional.

[16]The idea of functional presupposition was originally put forward by Bickhard (see for instance Bickhard 2000; Christensen and Bickhard 2002). We can understand the idea through the following example: "As everybody knows, the function of the heart is to pump blood, or more accurately to pump blood as a contribution to an ensemble of activities that result in blood circulation. The function that this serves, however, is to provide fluid transport for delivering nutrients to cells and removing metabolic end products. In this respect heart activity and cellular metabolism are interdependent processes. Without heart activity, fluid transport stops, and with it cellular metabolism. And if cellular metabolism ceases then heart activity also ceases, and subsequently fluid transport. In addition to heartbeat, cellular activity also produces other motor action that contributes to interaction processes such as breathing, food acquisition, eating and excreting. In turn these processes provide the resources required for cellular metabolism and expel waste products, thus contributing to the cellular processes that subserve them., . . . , These patterns of process interdependence in biological systems are (. . .) what determine the nature of organisms as viable (cohesive) systems" (Christensen and Bickhard 2002: 16–17).

The organisational account, therefore, interprets malfunction as any functional activity with respect to which there has been a *failure*[17] of regulation. In other terms, malfunctions are a subset of functions that fit first-order norms (of the first-order ongoing organisation in which they match functional presuppositions), but *not* second-order ones (since they do not obey to second-order regulatory functions, and prevent the shift to another first-order organisation). In this respect, the degree of malfunction of a trait could be assessed in terms of the set of first-order organisations of which it prevents the realisation. The degree of malfunction is, therefore, inversely proportional to the degree of adaptivity of the organism (see also chap. 4).

Malfunction occurs when the autonomous system fails in regulating the activity of a trait, including the specific case in which regulation aims at compensating for a "defective" activity of the trait in a given organisational regime. This is a crucial implication, because were a given first-order regime of self-maintenance capable of compensating for the apparently malfunctional activity of a trait, it would be impossible, from an autonomous perspective, to contend that the trait is malfunctioning. In such a case, its contribution to the system would, in principle, be indistinguishable from another contribution within the presupposed range of functioning. Indeed, if there were no higher-order regime with respect to which the behaviour of the trait is unfit, we would simply be faced, from the autonomous perspective, with a *different* organism (i.e. an organism that would function in a different, equally viable, way), and not with a malfunctioning one. For example, certain organs of the mole (i.e. those involved in sensory-motor activities) presuppose that its eyes provide very limited visual capacity, and that is why this animal is perfectly viable despite being almost blind (in fact some moles, like the star-nosed mole, display a remarkable foraging ability thanks to the star-shaped set of appendages that ring their nose). If there were no (failed) regulatory intervention, there would not be organisational criteria to interpret the behaviour of the trait as malfunctional. On the other hand, if regulation were able to compensate for the operations of a defective trait – by shifting to a regime of self-maintenance in which the trait would match the functional presuppositions – there would not be organisational reasons either to contend that the trait is malfunctioning. In such a case, its contribution to the system would match both first-order and second-order norms, and therefore it would be theoretically indistinguishable from any other contribution within the presupposed range of functioning.

A trait that malfunctions is, first of all, a functional trait, in the sense that it contributes to the maintenance of a self-maintaining organisation. What happens is that this contribution is not made *according to certain second-order norm* and that is why we say that it is a "bad" or "poor" contribution. Malfunctional traits show

---

[17]In technical terms, the very possibility to detect a failure of regulation supposes that the admissible ranges of the ongoing organisation and the alternative ones (to which regulation should move the system) are, at least *partly, non* overlapping. This means that the regulatory intervention must result in an observable *change* of the defective trait's activity.

a degree of malfunction rather than an "all-or-nothing", "function-no function" dichotomy. And the effects of a functional trait are deemed "good" or "bad" according to the norm that lies in the action of a regulatory subsystem (Saborido et al. 2014; Saborido and Moreno 2015).

It could be argued (Artiga 2011) that, ultimately, the norms to which any organism is subject have been set through evolution by natural selection, which shapes the species it belongs to. In particular, each given set of second-order norms could be defined at the populational level, because it has been selected in relation to the conditions of a stable existence (over a long period of time, covering many generations) in a given niche. Thus, it is because of its contribution to the self-maintenance of a class of organisms that this particular normative mechanism exists. And this happens too with the shaping of the structure and organisation of functional traits.

Yet, there are two aspects in which the current organisation of individual organisms matters. First, though the mechanism of adaptive regulation of a given organism is set through an historical-collective selective process – because only those forms of modulation that ensure viable organisations (in specific environments) can be selected – ultimately the regulatory mechanism would not exist if it did not make a contribution to the self-maintenance of each individual system in which it operates. The second, and even more important aspect is that although the *origin* of the norm according to which something is deemed malfunctional is ultimately an evolutionary matter, this does not mean that we cannot define, in the current organisation of each individual organism, whether or not a given trait is well-functioning or not. As Christensen has recently pointed out:

> The aetiologist may point out that, living systems have infrastructure for self-perpetuation largely as a result of an evolutionary history., ... nevertheless, ... the key perspective for normative evaluation of function is the current system rather than past selection. Regulation does not succeed by making parts function as they did in the past, it succeeds by making the system work well in present conditions. (Christensen 2012: 107)

In this respect, we consider that the organisational account of malfunctions can include evolutionary considerations without falling into epiphenomenalism, i.e., an understanding of functional attributions appealing to something other than the traits' current performance (see Sect. 3.1.2 above).

To conclude, we wish to emphasise that the organisational account to malfunctions does not rely on the subjective criteria of an external observer. What matters is what happens operationally within the system itself, and whether or not there is failure in adaptive regulation. Moreover, the normativity to which obeys the adaptive subsystem of a given organism is not defined with respect to a *type* of organisms but, rather, in relation to the current organisation of this organism and, more precisely, to the second-order closure to which regulatory functional traits are subject. This way of understanding the concept of malfunction is quite different from the most predominant notion of malfunction used in the philosophy of medicine, namely the bio-statistical conception, expounded by authors such as Boorse (1977, 1997), which claims that a malfunction is a deviation from "normal" (i.e., the statistically more common) functional behaviour. The bio-statistical conception has been fiercely criticised (Amundson 2000), and numerous problems and counterexamples have been put forward, so that its influence within the philosophy of medicine is declining (Khushf 2007).

The implications of an organisational account of malfunction are still to be explored and critically assessed. Yet, this account might open new directions in the search for a theoretical grounding of the notion of physiological disease, within an alternative naturalistic perspective.

4
Agency

The analysis of biological organisation conducted so far has focused on what we call its *constitutive* dimension. The autonomous perspective conceives the constitutive biological organisation as an emergent closure of constraints that, by grounding teleology and normativity, grounds functionality. As we emphasised, then, concepts such as organisation, closure, and functionality are inherently related and refer to a fundamental aspect of biological systems, which are able to maintain a network of interconnected functions by relying on the thermodynamic flow to which they are subject. Biological systems, first and foremost, (self-)maintain their coherence and identity as the closure between their constitutive constraints, which is also regulated by second-order constraints, so as to handle deleterious variations. Yet, the constitutive dimension of organisation is not *ipso facto* autonomous. In this chapter, we argue that autonomy involves also an *interactive* dimension, enabling biological systems to maintain themselves in an environment. We will refer to this interactive dimension as *agency.* A system that realises constitutive closure (metabolism) and agency, even in a minimal form, is an *autonomous system,* and therefore a *biological organism.* In this chapter, we will try to justify and develop this claim.

How does the autonomous perspective conceive the relations between the constitutive and interactive dimensions, between closure and agency? The general approach consists in viewing the constitutive and interactive dimensions of autonomy as being inherently related. As mentioned in the introduction to this book, autonomy is not independence, which means, in particular, that an autonomous system exists insofar as it maintains specific interactions with its surroundings, and therefore an adequate inward and outward flow of energy and matter. It is, then, the very coexistence of (and interplay between) the two causal regimes of autonomous systems that makes them agents: rooted in the thermodynamic flow, their organisation can only be agential.

Some of the ideas developed in this Chapter come from Moreno and Etxeberria (2005) and especially from Barandiaran and Moreno (2008).

A. Moreno, M. Mossio, *Biological Autonomy*, History, Philosophy and Theory of the Life Sciences 12, DOI 10.1007/978-94-017-9837-2_4

More technically, constitutive and interactive dimensions correspond to two nested classes of functions, both subject to closure. Accordingly, interactive capacities are a subset of constitutive functions that, as such, meet the definition of constraints given in Chap. 1. Interactive functions are, then, a set of constraints subject to closure, whose specificity lies in the fact that their effects are exerted on the boundary conditions of the whole system.

As an example of an interactive function, let us consider the movement of *Paramecium*, a eukaryotic unicellular organism. Paramecia have cilia, which are arranged in tightly spaced rows around the outside of the body. With the synchronised movement of these cilia, paramecia move into specific direction, so as to gather food (the *Paramecium* uses its cilia to sweep prey organisms, along with some water, through the oral groove, and into the mouth opening). The metabolic organisation of paramecia supports the activity of the cilia, which modify the environmental boundary conditions so as to propel the body of the organism in the direction of food. In turn, the ingestion of food contributes to the maintenance of the organism. Accordingly, cilia exert functional effects[1] that are subject to closure, just as any function is. Yet, as interactive functions, their specificity lies in the fact that the processes and reactions on which they exert a causal influence belong to the external environment (to the boundary conditions), which means, in turn, that these processes have not already been constrained by the biological system. In other words, the synchronised movement of cilia achieve an interactive function (an action) because it constitutes the first causal influence of a biological function on a given set of entities or processes (in this case, food and its position with respect to the organism).

As stated above, the conceptual distinction between constitutive and interactive functions is linked to their inherent and fundamental relationship. As a subset of constitutive functions, agential capacities are subject to closure and depend on the existence and stability of the whole biological organisation. In turn, they contribute to the maintenance of that very organisation by specifically managing its relations with its environment. As Kauffman (2000) has pointed out, autonomous systems, because of their agential capacities, can be said to "act on their own behalf": to determine themselves, living systems employ their functional constraints not only to maintain the constitutive organisation but also, and crucially, to promote its conditions of existence by modulating their surroundings.

Note, moreover, that the relationship between a biological organisation and its environment is asymmetrical: the organisation acts on the environment to

---

[1] Actually, cilia are very complicated structures. A cilium consists of a hollow, flexible cylinder, made from nine pairs of tiny tubes known as microtubules. Another pair of microtubules runs through the centre, connected to the surface by spokes. Each pair of microtubules has two protein molecules, known as dynein arms, attached to it at intervals along its length. These act like tiny motors, using adenosine triphosphate (ATP) as a source of energy. To achieve movement, they push in unison against the neighbouring microtubule pair, causing it to bend in the desired direction. So, in fact, the constraints responsible for movement are these proteins that selectively harness a set of physico-chemical processes leading to the whole movement of each cilium.

promote its own maintenance, while perturbations generated by the environment on the system are monitored in accordance with its own needs. The interaction is asymmetrical because it is guided by one side only, which imposes its own norms and aims on the other. Agency requires, therefore, the specification of an organisational core being the causal source (the "self") of the functional interaction. This makes the conceptual distinction, as we will discuss, between agents and other kinds of non-agential self-maintaining systems (in which they usually live) as, for instance, ecosystems.

The inherently interactive nature of biological autonomy, as well as the asymmetry in the interaction between the biological system and the environment are therefore the essential features of agency, as it is understood within the autonomous perspective. In what follows, we will develop the account in more detail.

4.1 What Is Agency?

Until recently, the concept of agency has received little attention from biologists and philosophers of biology.[2] One key reason for this neglect is that agency has usually been understood and discussed by cognitive science in connection to high-level human cognitive phenomena (such as beliefs, conscious intentions and reasoning, see for instance Davidson 1963; Dretske 1988), which, in turn, are difficult to handle within a naturalistic account. As Frankfurt (1978) has pointed out:

> There is a tendency among philosophers to discuss the nature of actions as though agency presupposes characteristics which cannot plausibly be attributed to members of species other than our own (Frankfurt 1978: 161–2).

This focus on human cognition is mainly due to the fact that human agents are supposed to be moved by "reasons" rather than "causes" which, in fact, is just another way of providing a non-teleological explanation of human behaviour (Davidson 1963: 691). Since most authors have traditionally agreed that satisfactory explanations should be non-teleological (unless rationalisation is considered a special form of causal explanation), only systems moved by reasons would be genuine agents.

For some years now, however, things have been changing. An increasing number of philosophers and theorists now acknowledge that the concept of agency can be pertinently applied to the biological domain, well beyond the specific case of human behaviour, and can include most biological organisms (see for instance Lyon 2006).

---

[2] As already mentioned in the Introduction, most of the authors we have included in the autonomous perspective, as Varela, Bickhard, Hooker, Christensen, Kauffman and Juarrero, have already considered that the question of agency is deeply linked to that of autonomy. Yet, only a few of them (Juarrero 1999; Skewes and Hooker 2009; Moreno and Etxeberria 2005; Barandiaran and Moreno 2008; Barandiaran et al. 2009; and more recently Shani 2012; Arnellos and Moreno in press) have developed an analysis of the very concept of agency, focusing in particular on how it can be originated.

One of the earliest philosophers to move in this direction was Frankfurt (1978) himself, who argued that non-conscious animals, like spiders, could be considered agents. More recently, Burge (2009, 2010) has defended the view that most living entities are agents.

In the domain of theoretical biology, the focus has predominantly been, to date, on the evolution of agency (which we will discuss in Chap. 7), while very little attention has been paid to the nature of agency itself.

Yet the extension of the notion of agency beyond its usual frontiers does not come without risks, and requires an adequate conceptual treatment. Indeed, such an "extended agency" would in fact be a highly simplified notion (it would notably lack rational and conscious intentionality), and yet would be assumed to share the fundamental features that are usually ascribed to agency, both in science and in common language. The aim is to specify the minimum requirements that a system must fulfil in terms of organisational complexity in order to be an agent and, accordingly, to obtain an adequate naturalisation of the concept in connection with biological sciences. At the same time, the focus on minimal agents would locate the debate about agency (and levels of agency) within a general framework, thus providing a way out of the maze of examples and counterexamples that are based on intuitive (and usually arbitrary) distinctions and definitions. If it is to make a relevant contribution, the autonomous perspective must be able to provide principled, non-arbitrary answers to questions of this kind.

What, then, is (minimal) agency, from the autonomous perspective? As mentioned in the previous section, agential capacities are a subset of biological functions that exert a causal effect on the environmental conditions of the system. This general characterisation has at least four main implications:

First, as recognised by both scientific and everyday language, a central feature of agents is their capacity to generate causal effects: agents are the source of interactions that are not determined by either the events of the immediate or distant past, or by physical laws of nature (Moreno and Etxeberria 2005). As Smithers (1997) pointed out, agents are systems that can initiate, sustain, and maintain an ongoing interaction with their environment, as an essential part of their normal functioning. A person blown away by the wind is not behaving as an agent, because he or she cannot be said to be the causal source of the movement. The autonomous perspective integrates this idea by identifying actions with a class of functions, and then with a class of constraints subject to closure. Accordingly, agents are the sources of causal effects because these effects are generated by the constraints that belong to their organisation.

Second, this characterisation of agency does not depend on the amount of energy invested, which might be higher in some cases, and lower in other cases. For example, fast swimming does require a relatively high investment of energy, while gliding flight is such that birds manage and constrain the flows of air to drive their flight in the desired direction, by means of minimal muscular contractions (Barandiaran et al. 2009). In the autonomous perspective, examples as gliding flight are genuine actions, just like those requiring a higher energy investment. As discussed in Chap. 1 (see Sect. 1.6), indeed, the energy invested by the system in performing a function is irrelevant to the definition of that function, which corresponds only to the constraining effect exerted on the boundary conditions. Accordingly, cases that might seem problematic for other accounts, such as bird gliding, comply perfectly with the theoretical requirements established by the autonomous account, and can therefore be taken as genuine actions for principled reasons.

Third, actions are performed according to a certain goal or norm. In contrast to mere "effects" of the system on the environment, actions are supposed to have goals and comply with norms. Actions have teleological and normative dimensions. Again, the autonomous perspective captures these requirements by deducing them, in a principled way, from the fact that actions are a class of functions. As such, they are subject to closure, which implies, in particular, that they possess the teleological and normative dimensions generated by the realisation of closure, as we explain at length in Chap. 3. By contributing to the maintenance of the closed organisation to which they belong, agential functions contribute to maintaining the conditions of their own existence; hence, the maintenance of the whole organisation can be taken as the naturalised goal of agential functions, and its conditions of existence are the norms of their activity. A bacterium moves its flagella and approaches a concentration of sugar: these interactions contribute to the maintenance of the bacterium itself, and the conditions of the maintenance constitute the norms against which the movement can be evaluated.

Fourth, just like any class of functions, agency requires the interplay of mutually dependent constraints, each of which makes a different yet complementary contribution to the maintenance of the whole system. In other words, it must be possible to discriminate between agential functions and other classes of functions at work in the system: a system in which this distinction cannot be drawn is not an agent. For instance, in spite of the fact that they interact with the environment and realise self-constraint, entities such as riverbeds or hurricanes are not agents, precisely because they exert a single constraint on their own boundary condition. Agential capacities are functions, and to ascribe functions one must distinguish between different causal roles in the system, a division of labour among the parts, so to speak (see Chap. 3, Sect. 3.2.2 above on this issue). And, again, this is what happens when closure of constraints is realised, which is why functions are not ascribed to dissipative structures.

In the specific case of agential capacities then, the autonomous perspective maintains that a system is an agent only if in addition to exerting constraints on its environment in a teleological and normative way, its organisation also includes different *classes* of functions, and cannot be reduced only to the interactive dimension. In particular, the causal source of the agent's interactive capacities should be an integrated organisation being at the same time conceptually distinct from and, in the end, dependent on these interactive functions it generates. In other words, a system is an agent only if it possesses a form of individuality that cannot be reduced to the fact that it interacts with its environment.

With these implications in mind, let us now turn to minimal agency. In the following section, we will try to provide some keys to understanding how agential capacities can be realised in biological systems, in their minimal forms.

4.2 Minimal Agency

In Chap. 1, we suggested that the constitutive organisation of biological systems realises closure, and specifically closure with regulatory capacities. We labelled this kind of organisation *metabolism*. Yet, as discussed above, as autonomous systems, biological systems possess an inherent interactive dimension, in the form of agential capacities, whose general aspects have been expounded in the previous section. In this section, we will examine what form agential capacities may take when realised by minimal metabolisms: we will therefore deal with *minimal agency*, i.e. agency as expressed by minimal biological organisations.

When addressing the question of minimal agency from the autonomous perspective, two unsatisfactory options – both advocated in the literature – should be avoided. Firstly, some proposals situate minimal agency in too complex systems, and set too demanding (and, we hold, unnecessary) requirements. In this case, minimal agency is not really minimal, and more simple systems could still be justifiably taken as genuinely agential. And secondly, other proposals interpret too simple systems as agents. Here, in contrast, the weakness is that the target systems do not satisfy the requirements for being considered agents in a relevant sense.

Most of the advocates of the first option are philosophers who, as mentioned in the previous section, have traditionally discussed agency in connection with high-level cognitive capacities. During recent decades, animal agency has received an increasing amount of attention, although the question of minimal agency has not yet been addressed in explicitly conceptual terms. As a matter of fact, most of the examples typically discussed include animals with nervous systems, which implicitly seems to indicate that minimal agency tends to be associated with metazoans. As mentioned, one exception in this respect is Burge (2009, 2010), who has recently formulated in explicit terms the question of what criteria would be pertinent to define minimal agency. This author believes that agency only occurs in the case of a functional interaction triggered by a living system as a whole, and implying movement or behaviour; therefore any other type of functional interaction – say, an adaptive secretory process in response to a stimulus – would not be taken as an action. In his view:

> [various kinds of taxes] even in very simple organisms, are instances of primitive agency. The paramecium's swimming through the beating of its cilia, in a coordinated way, and perhaps its initial reversal of direction, count as agency (Burge 2009: 259).

At the same time, he discards the case of lower organisms, like bacteria, arguing that:

> True taxes in prokaryotes are rare or absent, because the small size of the prokaryotic cells does not admit of much diversity on the cell body or of sufficient capacity to register the small differences that must be differentiated (Burge 2009: ibid.)

The main problem with Burge's own account is that it does not seem to provide an explicit criterion for justifying the exclusion of prokaryote motility, while including paramecia motility. Actually, Burge's tentative definition of minimal agency:

> The relevant notion of action is grounded in functioning, coordinated behaviour by the whole organism, issuing from the individual's central behavioural capacities, not purely from sub-systems (Burge 2009: 260)

seems to rely more on intuition than on a rigorous conceptual base. Although it points in the right direction, in our view, Burge's account does not fully succeed in establishing explicit, adequate criteria for characterising minimal agency, and it seems that the concept could be applied to phenomena that are simpler than those he has in mind.

The second strategy is that adopted by the "ultra-minimalist" proposals, usually developed in the literature about the origins of life. A representative example is the model of the self-propelling oil droplets (Hanczyc and Ikegami 2010), in which a combination of chemical reactions, self-assembly processes and convective phenomena triggers the spontaneous global movement of an oily system in an aquatic environment. According to these authors (Horibe et al. 2011), self-propelling oil droplets can be taken as "models of autonomy and minimal cognition based on physicochemical principles" (Horibe et al. 2011: 718). In the same research line, much more complex chemical "nanorobots" have been recently developed (Sengupta et al. 2012; Lagzi 2013)

In a similar ultra-minimalist vein, but with very different assumptions about what matters in our search to understand the origins of life, several other authors have proposed that minimal agency took place in the form of self-replicating molecules or "replicators".[3] According to this view, replicators "perform actions", since they select from the molecular environment the "building blocks" required for their own continuous replication. Dennett (1986), Chap. 2), for instance, claims that a minimal agent is a Darwinian system, blindly generated by natural selection and possessing different hardwired phenotypes. The responses of these systems to survival problems are determined by their genetic inheritance and are quite inflexible. But Dennett nevertheless understands such "minimal Darwinian agents" to be self-replicating molecules.

In this line of argument, the case of viruses and prions is worth mentioning as relevant candidates as minimal agents, because in addition to being replicators, these systems (especially viruses) also perform other "functions" when infecting living cells. First, viruses infect the host cell and force them to fabricate copies of the virus, while prions act as a template to guide the unfolding of more proteins into prion form. In turn, these newly-formed prions go on to convert more proteins, thus triggering a chain reaction that produces new copies of the original prion. But viruses, in addition, are quite complex molecular systems, constituted by several catalytically-active macromolecules, whose secondary structures are organised in rigid parts and which may display relative movements, thus generating

[3]Dawkins proposed the idea of the "replicator", defined as "anything in the universe of which copies are made" (Dawkins 1982: 83). In the context of the origin of Life, "replicators" are the initial molecules that first managed to reproduce themselves and thus gained an advantage over other molecules within the primordial soup.

different interactive effects. These effects may include small displacements or other mechanical effects described by a list of terms borrowed from the description of machines: lever and spring, ratchet and clamp, etc. In some cases, the interactions are triggered by the potential energy of the virus itself. For example, viruses attach to the host membrane and, traversing the host cell wall, inject their DNA inside the cell. In other cases, the interaction also requires some external supply of energy, which is provided by the machinery of the infected cell. Because of these interactive capacities and because apparently they "act" on their own behalf – some authors maintain – viruses (and prions) can be taken as minimal agents.

Do entities such as droplets, chemical nanorobots, or replicators such as viruses and prions meet the requirements for agency listed in the previous section? Are they genuine minimal agents, as some authors claim?

From the autonomous perspective, this conclusion is not compelling. The examples are interesting, because they show that, in some cases, biological-like behaviours can in fact be performed by systems endowed with low (or very low) organisational complexity. Yet, we hold that none of these systems meet the requirements established by the autonomous perspective, and cannot therefore be taken as minimal agents. One fundamental reason is that most of these entities do not possess a constitutive organisation that is complex enough to perform agential capacities: they exhibit such capacities only insofar as they are integrated into much more complex systems (typically: cells) that *are* organised, in the specific sense that they realise a closure of constraints. In a hypothetical prebiotic scenario, in which these entities are isolated and not incorporated into cells, they would not exhibit agential functions. Even though some of these entities (such as Hanczyc and Ikegami's droplets, or the more complex chemical nanorobots), taken in isolation, may do something, they would not meet the requirements established to be taken as organised systems and, *a fortiori*, as agents. Self-propelling oil droplets and nanorobots do certainly move, but these systems lack a metabolism allowing them to display an inherent capacity to modify the conditions at the system-environment interface.

This is also the case of prebiotic replicators, which might have been able to make copies of themselves before the existence of proto-metabolic organisations. For some, like Pross (2008), replicators represent the essence of agency, because they achieve an "inversion of kinetic and thermodynamic directives", which the author identifies as "purposeful". Admittedly, Pross does not claim that a nude molecular replicator is already an agent, only that replication is the basis of the teleological nature of life (and therefore, of minimal agency). However, even acknowledging that replication may drive chemical systems away from thermodynamic equilibrium, the teleology or purposefulness of agency requires something else. Claiming that a replicator is an agent because it achieves a kind of causal loop, such as self-replication by template, does not fit with our fourth requirement. Hence, self-replicating structures cannot be considered agents from the autonomous perspective since, as explained above, agency requires a constitutive organisation which includes different classes of functions; at the most, if we consider replicators associated with catalysts and so performing replicative cycles, they can be reduced to one single capacity: replication. In this sense, they are similar to dissipative structures that can *only* perform self-maintenance. Accordingly, they are not agents.

Arguably, more complex replicators, such as viruses, could do many other things, but only because they recruit the machinery of the host cell. Hence, it is the cell modified by the virus that can be justifiably taken as an agent, not the virus itself. As Spiegelman's experiment (Kacian et al. 1972, see Chap. 5) demonstrated, it is the complex organisational closure of the cell, insofar as it provides a rich functional diversity, that enables the virus not only to replicate but also to perform other functions. Therefore, viruses are not agents.

The autonomous perspective, then, provides explicit criteria to guide the search for minimal agency, and to avoid the inclusion of both too complex and too simple examples. Since autonomy is not independence, closure requires agency as a set of constraints devoted to handling interactions with the external environment in accordance with intrinsic teleology and normativity.

The next question, then, is: what kinds of system would fit the definition of minimal agents from the autonomous perspective? Let us recapitulate on what we have established so far. The autonomous perspective inherently links agency to organisational closure, and provides explicit criteria for ascribing agential capacities to a natural system. In particular, we submit that it allows minimal agency to be characterised in a principled way, and excluding both too complex and too simple cases. A minimal agent is a system fulfilling the four requirements formulated in the previous section: condition three requires that the system have an intrinsic normativity; and conditions one, two, and four require that the system exhibit differentiated constraints. In particular, condition four requires that in order to be an agent, a(n organisationally closed) system should exert constraints on its own boundary conditions; in other words, it should include at least one constraint subject to closure which acts on the boundary conditions of the system.

As an example of minimal form of agency in present day life, let us consider the case of taxes in bacteria, already evoked at the beginning of this chapter. A "taxis" is a movement of an organism triggered by a given feature of the environment, whose presence has some relevance for its self-maintenance. The movement could be either towards or away from that feature: in both cases, the taxis is driven by the organism, which employs energy to generate it. Actually, there are a wide variety of taxes, depending on the type of feature in the environment – barotaxis (pressure), galvanotaxis (electricity), phototaxis (light), magnetotaxis (magnetic field), thermotaxis (temperature changes) – although the most common one is chemotaxis, triggered by some specific chemical concentration gradient.

A well-known example of chemotaxis is that performed by the bacterium *E. coli*. When a certain concentration of sugar is detected in the environment, *E. coli* is equipped with flagella that drive it towards the concentration gradient (Neihardt 1996). The movement is the result of the coordination between membrane receptors and motor mechanisms, mediated by metabolic pathways in the cell (Losik and Kaiser 1997; Hoffmeyer 1998). In particular, some proteins located on the membrane detect sugar molecules and trigger metabolic pathways, which change

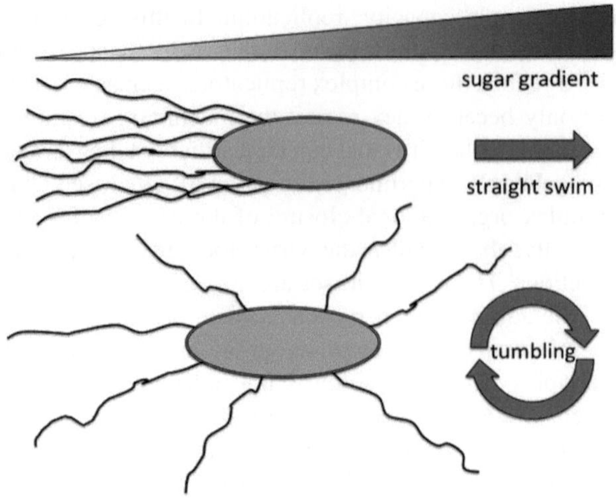

Fig. 4.1 Two different movements of bacteria depending on the presence of sugar in the environment (*credits: Juli Peretó*)

the movement of its flagella; in turn flagella generate the swimming towards the sugar gradient (instead of the usual tumbling movement) (Fig. 4.1).

We submit that the interaction induced through chemotaxis, however minimal, satisfies the four requirements for agency. On the one hand, chemotaxis is not a spontaneous pattern, and has therefore an energetic cost for the organism (it consumes ATP molecules). The cell, through its flagella, triggers a movement that constrains its boundary conditions and significantly change them. In particular, it canalises the displacements of the body and locates it in a different environment, where the concentration of sugar is higher. On the other hand, such a high concentration will enable the organism to absorb the sugar and, thereby, to contribute to its own self-maintenance.

Chemotaxis is therefore a causal effect subject to closure, which contributes to the maintenance of the cell. More specifically, we submit that chemotaxis is an interactive function because it constitutes the first causal influence of a biological function (exerted by the flagella) on a given set of entities or processes (in this case, the nutrients and their position). Chemotaxis constitutes a minimal form of agential function and metabolisms endowed with this function are (minimal) agents.

4.3 Adaptive Agency

The essence of agency lies in the capacity of an autonomous system to functionally constrain interactive processes, and thereby ensure its own maintenance. Of course, however, agents exert actions not only in stable environments, but also in all those situations in which environmental conditions vary, because of some sort of external perturbation, in a way that can be deleterious. As we discussed in Chap. 1, Sect. 1.8, biological systems are organisationally closed systems that possess adaptive regulatory mechanisms enabling them to cope with (some of) these variations, be they internally or externally generated. In this section, we focus specifically on those regulatory responses to changes in external conditions which induce a change of agential functions: the system reacts to external changes that threaten the viability of the system by modifying its interactive behaviour in a way that ensures its viability. Systems endowed with these specific regulatory capacities are *adaptive agents*.

As a matter of fact, virtually all present-day organisms – even the simplest ones – possess the capacity to adapt their actions in somatic time, in accordance with different environmental conditions. Exceptions include certain types of bacteria that live in very homogeneous and stable environments, such as, for instance *Buchnera* or *Wigglesworthia,* which are endosymbiotic bacteria living in the cells of certain insects. These endosymbionts, which are at the frontier between organisms and cellular organelles, benefit from reduced exposure to predators and competition from other bacterial species, as well as from the ample supply of nutrients and relative environmental stability inside the host, and have lost many of the adaptive functional capacities that their ancestors possessed when they were free-living bacteria. These biological systems regulate their constitutive functions, but regulation in this case does not involve agency.

Apart from these cases, however, all biological systems are adaptive agents. They are able to detect potentially deleterious variations in the environment and to trigger the selection of an adequate functional action from within the available repertoire.

As for general regulation, adaptive agency was probably preceded by more primitive forms of agency. Presumably, cellular proto-metabolisms were homeostatic, in the sense that they compensated for internal and external perturbations by means of feedback mechanisms integrated and distributed into/around their constitutive organisation; basically what we call "constitutive stability" (see Chap. 1, Sect 1.8). However, when these prebiotic systems increased in complexity, they became more fragile: noise and environmental perturbations affected their organisation, which was, given its holistic nature, easily disintegrated. This bottleneck was overcome when proto-metabolic systems began incorporating a regulatory mechanism capable of exerting an active control on its interactive (and constructive) processes, detecting different conditions and monitoring its own constitutive processes so as to avoid or prevent deleterious situations. Adaptive agency, therefore, entails a capacity for detecting relevant changes and selectively triggering functional processes before the system disintegrates (Barandiaran and Moreno 2008).

The point is that adaptive agents react in a highly novel way: they do things according to what they detect. The environment is not only a source of indistinguishable perturbations, but also of specific, recognisable ones. As discussed in Chap. 1, Sect. 1.8, these (recognisable) perturbations trigger the regulatory system but do not directly determine the system's response (because the constitutive regime is modified by the regulatory system, not by the perturbation itself). In turn, although the modification of the constitutive regime does not affect the regulatory system, it can affect the perturbation (the organism eats the new food, or secretes chemicals to neutralise a lethal substance). Due to the action of the regulatory system, the constitutive regime is modified so as to fit in with a specific perturbation.

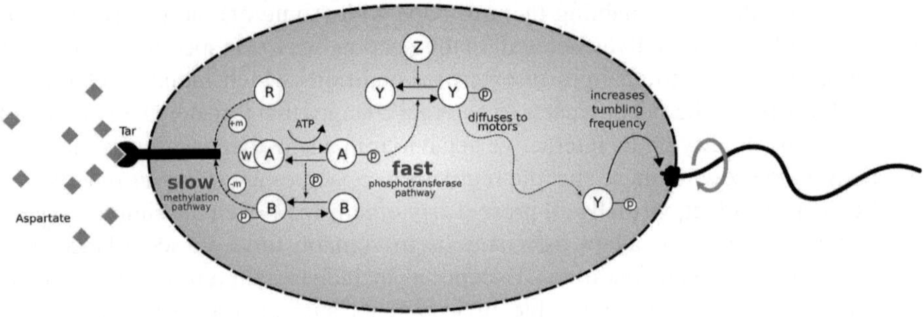

Fig. 4.2 A detailed model of the mechanism supporting chemotaxis in the bacteria Escherichia Coli (Source: Barandiaran (2008). This file is licensed under the terms of the GFDL (GNU Free Documentation License: available at https://www.gnu.org/licenses/fdl.html). Retrieved from http://www.sindominio.net/~xabier/phdthesis/)

Interestingly, the specific perturbation becomes a "recognisable stimulus" (Bich et al. forthcoming) because of the nature of the relationship existing between it and the regulatory system: the regulatory subsystem is sensitive to the stimulus, in the sense that it establishes specific classes of equivalence with respect to these specific perturbations (Rosen 1978) and the way they are activated.

For example, let us consider again chemotaxis. Swimming and tumbling are two different interactions that are both appropriate in the sense that they contribute to self-maintenance in differing conditions, but the bacterium can usually switch between them as the conditions change. In other words, when bacteria detect that their basic constitutive organisation is approaching disruption (in particular, when their metabolic maintenance is at risk due to lack of sugar), the regulatory mechanism induces changes in the movement of the flagellum (if the presence of this product is detected) so as to achieve the environmental conditions necessary (by approaching a region of higher concentrations in sugar) to bring the system back to a viable situation (Fig. 4.2).[4]

[4]In *E. coli*, responses are mediated by the phosphorylated response-regulator protein P-CheY. Signals are passed from the receptors to cytoplasmic chemotaxis components: CheA, CheW, CheZ, CheR and CheB. These proteins regulate the level of phosphorylation of a response regulator called CheY that interacts with the flagellar "motor switch complex" to regulate swimming behaviour. The sensitivity of the chemotaxis system depends on allosteric nonlinear effects within the chemotaxis signal transduction network (Alon et al. 1998). Despite the complexity of the details, the essence is that the stimulus I changes an allosteric molecule R and alters its properties. In the absence of the stimulus, R connects to the constraint C by enabling the production of branch A; in the presence of I, R modifies the activity of C by enabling the activation of branch B. While in the absence of regulation the stimulus favours the functioning of a pre-existing pathway/attractor to the rate allowed by the structure of the constitutive regime, in this case, the regulatory mechanism transforms a rate difference into an all or nothing difference. The system is now able to cope with stimulus I so that the effect of the perturbation is not just compensated for, but actually integrated in the organisation of the system by bringing its effect inside. The regulatory loop closes because the system (through the regulatory transition) becomes able to interact viably with stimulus I.

As Barandiaran and Egbert (2013: 29) have pointed out, in bacterial chemotaxis

> it is assumed that natural selection has tuned sensor transduction mechanisms, sensorimotor chemical pathways and flagellar rotation speed and probabilities so that behaviour turns out to be adaptive (...) In this way the bacterium is capable of modulating its behaviour in direct causal correlation with its viability dynamics and not just by responding to external conditions.

In short, in order to be considered an adaptive agent, a system must satisfy two conditions. Firstly, it must be able to constrain its boundary conditions to ensure its self-maintenance. This in turn requires that it possess an internal organisation that is the causal source of the interactive processes ensuring its identity. Secondly, it must be able to discriminate between specific processes or structures in its environment, and to functionally act on them.[5]

4.3.1 The Specificity of Motility

Even the simplest organisms – prokaryotes – are capable of displaying a huge variety of adaptive strategies in their environment. They can induce metabolic changes according to differentiated environmental conditions (i.e., the adaptive mechanism of the lac operon discussed in Chap. 1); they can control their position and move towards better environmental conditions; and they can induce morphological and structural changes to form colonies when this collective organisation increases their chances of survival (Young 2006). Bacteria can also exchange genes (such as genes enabling bacteria to develop antibiotic resistance) between members of mixed species colonies (Miller 1998). Bacteria use "quorum sensing" to assess their own population densities and modify their behaviours accordingly (Miller and Bassler 2001). Quorum sensing plays an important role in regulating intercellular signalling in accordance with cell population density, so as to enhance beneficial cooperative behaviours. Some of them can even drastically slow down their metabolic activity and transform into spores when the environmental circumstances become too dry or too hot. This vast array of adaptive actions is accomplished through functional modifications to these organisms' plastic metabolism, tuned to relevant environmental changes.

[5]This has prompted some researchers (Weber and Varela 2002; Thompson 2007) to claim that adaptive autonomous agents are able to "enact a meaningful world" and that (since they are capable of generating their own individuality and regulating it according to norms) these systems "make sense" out of their functional/dysfunctional interactions, i.e., these interactions or encounters acquire "significance" and "valence". Despite the apparent and intuitive sense of these claims, their scientific adequacy is disputable. For, if we characterise these interactions as functional when they contribute to the maintenance of the agent (or as dysfunctional when they undermine it), what do we add with this rather anthropological terminology? As we shall see in the Chap. 7, this is a very controversial issue. See also Barandiaran (2008) for a critical analysis.

All forms of adaptive agency in bacteria require a relatively complex organisation a metabolism regulated in somatic time by a conservative structure (DNA) or by a different and relatively independent subsystem of chemical reactions (Van Dujin et al. 2006). As Bonner (2000) has pointed out, the evolution of molecular adaptive mechanisms, such as configurable membrane proteins coupled to processes that rapidly adjust and regulate gene-expression and metabolism, allowed organisms to better adapt to rapidly changing environmental conditions. These regulatory systems may have partly consisted of locally acting regulator genes that were responsive to very specific environmental features (Van Dujin et al. 2006).

Thus, a very simple form of adaptive agency is achieved through the functional control of the genome, whose expression is regulated by the system according to environmental conditions.[6] Since part of the genome is composed of a set of metabolically decoupled[7] gene strings, the adaptive mechanism consists of the activation and deactivation of genes in order to switch between metabolic pathways in accordance with certain environmental conditions. Agency here takes rather the form of self-transformation than of a direct modification of environmental conditions. A similar case is that of some bacteria that, under harsh conditions, transform into spores in order to better resist heat and dehydration.

However, in other cases, adaptive agency does not directly involve genome-specific activation. For example, adaptive agency takes place when a whole subsystem of biochemical pathways not involved in the constitutive metabolic network supports detection-action coordination. In this case, regulatory chemical pathways act directly on the interaction between the system and the environment.

The example of chemotaxis in *E. coli*, mentioned several times above, is of particular interest because it is precisely a case in which adaptive agency takes place through motility, which constitutes a minimal form of "behaviour". Motility is an agent's capacity to move under its own power, so that it is able to perform fast (relative to its size) directional movements aimed at changing its environment in search of more favourable conditions. The detection of (and functional response to) relevant environmental changes constitute, in the case of adaptive motility, a sensorimotor cycle, whose viability is strongly affected by size-time limitations. It is this high size-time (speed) limitation that distinguishes sensorimotor adaptivity from other forms of adaptivity (Moreno and Exteberria 2005).[8] The appearance

[6]In the course of evolution more complex forms of these genome-based regulatory control systems appeared, permitting regulatory genes to exert increasingly global control over metabolic functions, thereby becoming sensitive to new external features (Lengeler 2000).

[7]See Chap. 1, Sect. 1.8 for a discussion of the "decoupling" of regulatory mechanisms from constitutive organisation.

[8]However, in primitive life forms the metabolic subsystem supporting motility is not substantially different from that supporting other forms of adaptability. For example, when the prokaryote *Caulobacter* finds itself in a very humid medium it remains fixed to the soil like a vegetal type, whereas, in dry periods, it reproduces and the new cells grow a flagellum capable of transporting them to a more humid environment. So, the interactive loops established by the most primitive organisms with their environment are always contrasted and evaluated according to the

of behavioural adaptive agency implies not only the linkage of two processes, (the maintenance of constitutive closure through recursive interactions with the environment and the maintenance of this interactive cycle in accordance with its effects on constitutive closure) but more importantly, this intertwined regulation is strongly affected by size-time limitations. One very interesting consequence of these size-time limitations is that, as size increases, those biological organisations that support agency through biochemical mechanisms become severely restricted in their capacity for displaying efficient behavioural agency (Barandiaran and Moreno 2008).

Of all the forms of adaptive agency, and for the reasons stated above, motility is therefore of special interest. When we consider organisms whose way of life is based on motility, a bottleneck appears for several reasons. Firstly, the level of complexity that the adaptive subsystem can achieve (within the biochemical medium), without severely interfering with metabolic processes, is very limited. Secondly, as the size of the organism increases, the fast and plastic correlation between sensor and effector surfaces becomes more difficult (or even impossible in multicellular organisms), because of the slow velocity of diffusion processes. And thirdly, the organism must solve the problem of achieving unified body coordination for displacement. The type of rather sophisticated motile agency displayed by *Paramecia* illustrates the tension produced by the combination of these three factors: epithelial conduction (through Ca channels) is used to enable fast and coordinated beating of cilia because, unlike in the case of *E. coli,* this could not be achieved by mere diffusive mechanisms. However, the complexity (in terms of functional diversity and integration) that homogeneously spreading epithelial conduction can achieve is severely limited (Barandiaran and Moreno 2008; Keijzer et al. 2013).

The appearance of multicellularity posed an important challenge in the evolution of this form of agency, since at that size, biochemical mechanisms cannot support fast and versatile motility (Moreno and Lasa 2003). There are two reasons for this: the greater internal distance between parts of the body, which need to be connected in very short spaces of time (so that the organism can move quickly and in a coordinated manner); and the need to modulate the organisation of connections selectively (to achieve the adequate sensorimotor correlations) for versatile, plastic agency. Hence, if biochemical network plasticity were the only mechanism for accomplishing adaptive interaction and self-maintenance, the forms of movement-based agency would probably be very limited at the multicellular size.[9]

---

effects they have upon their basic capacity for self-construction (or self-maintenance), which is their main normative goal. In fact, in prokaryotic cells, body movement could be considered as simply an extension of the set of mechanisms that are required for a minimal metabolism. Thus, capturing food by means of body movement (as opposed to exploitation of primary energy resources or fermentation processes) does not entail qualitatively important differences in adaptation mechanisms. At this level, the underlying organisation of behaviour and morphological change is basically the same.

[9]It has been argued that plants possess epithelial cells, which can be sensitive to local chemical or tactile stimuli, triggering a change of electric potential capable, in principle, of producing

As we will show in Chap. 7, the bottleneck was only overcome by the appearance, in the evolution of some kinds of multicellular organisms (metazoan), of a new kind of cell capable of forming a tissue, the nervous system, along with a whole set of bodily changes that gave rise to qualitatively different, much more complex, forms of agency.

4.4 Autonomy

Having established what an agent is from the autonomous perspective, let us now turn to the general theme of this book, biological autonomy. We submit that agents, i.e., in an extended form, systems realising a *regulated closed agential emergent organisation*, in the technical sense developed throughout these four chapters, are *autonomous* systems and therefore biological *organisms.* Any natural system lacking one of the above features would be, by definition, *infra*-biological.

What is the logic behind this definition? Its central purpose consists in showing that autonomy is inherently grounded in, *and yet not equivalent to*, organisational closure. As we claimed before, the central feature of biological organisms, understood in terms of autonomy, is their capacity of self-determination, the fact that they "are what they do". Closure, described in Chap. 2 as an emergent regime of causation, is the technical concept that captures this capacity. Yet, closure is not autonomy: the inherent complexity of biological systems requires also appealing to regulation and agency.

As we argued at the end of Chap. 1, Sect. 1.8., we take regulation as a necessary condition for autonomy because it confers a conceptually stronger sense to self-determination. Regulated closed organisations not only generate intrinsic norms but, in addition, modulate these norms in order to promote its own maintenance, and this in accordance with second-order norms. Auto-nomy here is not just the maintenance of the current condition of existence, but the fact of promoting its own existence on behalf of a more fundamental (and less contingent) identity. Moreover, a closed network without regulation cannot harbour an open domain of functional diversity (just the opposite in fact, since, in practice, its functional

fast agential responses (Simons 1981). But plants' intercellular communication is not based on epithelial cell communication, which lacks directional and selective propagation and is unable to organise the modulation and regeneration of signals. Instead, the communication system of plants is based on channels called plasmodesmata, which work by transporting (either passively or actively) a large variety of chemical signals. However this mechanism is a far cry from showing the speed, plasticity and recursive modulation of signals of neural networks. Not surprisingly, then, plasmodesmatal connections seem to be limited to adjacent cells (Trewavas 2003). Moreover, the body plan of plants does not allow them to develop musculoskeletal structures, which by virtue of their ability to channel energy into reversible mechanical motion, are of fundamental importance for behavioural agency. For a detailed discussion on the limitations of plant's agency, see Arnellos and Moreno, in press.

diversity is very limited), and cannot therefore provide the organisational core for biological evolution. Similarly, as discussed above, in the absence of regulation no proper agency can take place. Autonomy hence also requires agency, as we have been arguing throughout this chapter. The very nature of biological organisation, grounded in the thermodynamic flow, is such that it must control the interactions with the environment and, in particular, ensure an adequate inward and outward circulation of energy and matter. The openness of autonomous systems requires (possibly adaptive) agency.

It is very important to underline that, at this point, we have a characterisation of *minimal* autonomy. This term describes the "organisational core" of all biological organisms, even in their minimal realisations and expressions, although in no way does it claim to cover the amazing variety and complexity of forms that biological organisation can take. In the following chapters (particularly Chap. 6), we will offer some ideas as to how the autonomous perspective may provide useful tools for understanding more complex biological phenomena. However, minimal autonomy does capture the essential features of biological systems, and therefore, we believe, the concept of *organism* itself. In particular, minimal autonomy can be pertinently applied to account for the principles of organisation of *unicellular* organisms.

However embedded in evolutionary and ecological webs organisms might be, both their metabolic functioning and their agency rely on the fundamental organisational core characterised so far. This is an important issue because without a highly integrated and cohesive individual organisation as captured by the concept of minimal autonomy, living systems would not possess the necessary requirements for their long-term maintenance and evolution, nor would they be able to build broader and more complex organisations. Furthermore, without a theoretically well-founded definition of minimal autonomy, it would be very difficult to provide an organisational grounding of metabolic organisation, individuality, agency, unit of selection, etc., or to make useful distinctions between organisms and other forms of cooperative or "ecological" networks (Ruiz-Mirazo and Moreno 2012). After all, increasingly complex forms of individual agents have emerged and developed, bringing forth progressively more sophisticated constitutive properties and interactive capacities. In the following section, we make a first step in this world of higher-level biological organisations.

4.5 Beyond Individuals: Networks of Autonomous Agents[10]

Adaptive agents functionally modify their environments for their own sake. But, in the same way, they can also modify the organisation of other adaptive agents. Therefore, a very important consequence of the agential dimension of autonomous systems is their capacity to establish complex webs of interactions amongst themselves, with these interactions possibly giving rise to new higher-level functions.

[10]Many of the ideas developed in this section are taken from Nunes et al. (2014).

These webs of interactions can be of very different types: some lead to the (more or less temporary) constitution of relatively cohesive collective aggregates (colonies, symbionts, societies, swarms, flocks, multicellular organisms) and involve essentially strong dependencies between organisms; others are more basic and concern weak relations, based on metabolic complementarities between different types of organisms (in terms of how they obtain their energetic resources), namely, the organisation of the movement of materials and energy through living communities. These latter ones are known as *ecological* interactions.

Strong dependencies between autonomous systems essentially involve specific structural or behavioural changes in the resulting higher-level system. This is the case, for example, with intercellular signals in a colony or in a process of development, or similarly, with social coordination. We shall discuss these types of interactions in Chap. 6. For their part, weak interactions refer to cases of minimal constraints exerted by an organism on its surrounding physicochemical environment, in a way that influences the environmental conditions of another (group of) organism(s). In this last section, we shall briefly discuss the nature and role of these ecological interactions. We have chosen to discuss weak interactions in the first place because they constitute the first level of complexification of life at higher levels of organisation.

We also want to discuss ecological networks here because they are a relevant case in which some of the key concepts of the autonomous perspective (in particular, those of organisational closure and function) can be applied to systems whose constituents are themselves organisms, and therefore autonomous agents, interacting in a functional way with their surroundings. Here, there is a qualitative shift in the organisational scale, which has significant implications for understanding the higher-level system itself.

Ecological interactions can be produced by both the most complex types of agents (say, a jaguar hunting its prey) and the simplest ones (as an autotrophic bacterium providing organic matter for worms in a deep sea vent). Very likely, hence, ecological interactions appeared early in the history of life precisely because they can result even from the most basic forms of agency. Different groups of early organisms, having developed an increasing ability to modify their environment, would have adapted their forms of life, creating complementary relations, which are the basis for building ecological systems (Fig. 4.3).

What matters in ecological interactions, from the autonomous perspective, is the fact that a given action performed by a specific type of organism on a specific environment affects the energy and material inflow of another type of organism, which in turn performs an action affecting another group of organisms and so on, until the network of interactions folds up. The importance of this closed network of ecological interactions lies in the fact that they allow long-term sustainability of life in both energetic and material terms.[11] The action of each type of organisms ensures

[11]Of course, the question of the long-term sustainability of life has another very important dimension, namely, how such a complex organisation can be maintained (and eventually increased) through intergenerational changes. We shall discuss this fundamental question in Chap. 5.

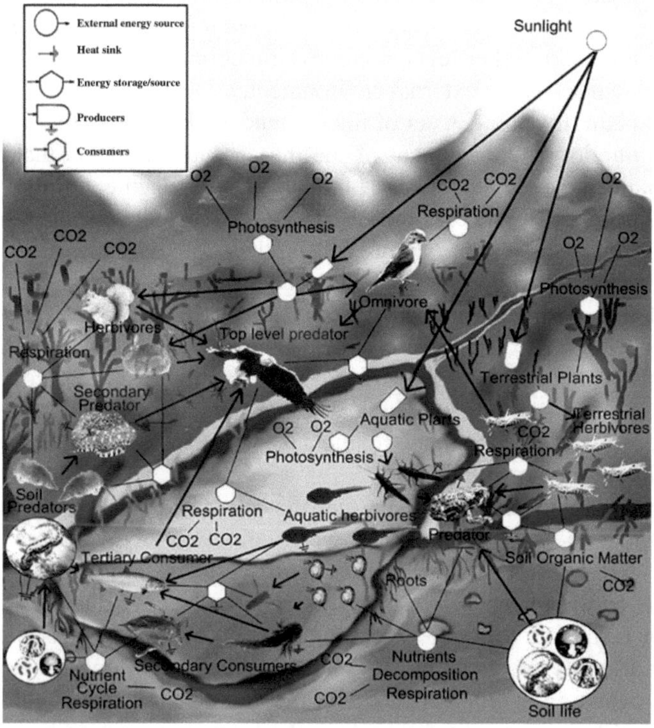

Fig. 4.3 A schematic model of a system of ecological relations (Source: Wikimedia Commons, Mark David Thompson)

that the flux of energy and matter necessary for its own maintenance, as well as that of the other types of organisms, is constrained so as to be indefinitely maintained (provided that certain geological and astronomical conditions are met: for instance, the network is ultimately driven by a stable external energy source, such as the sun).

Ecological systems are therefore a kind of "biologically constructed environment" (Dagg 2003), which can be characterised as higher-level closed organisations: the functional units are the different groups of organisms, which depend on each other because of the way each one constrains the ecological flow of matter and energy. Yet, these actions become ecologically relevant and functional only when they pass a certain threshold (rate of production) so to have causal consequences on the conditions of existence of other groups, and hence to contribute to the maintenance of the whole network of interactions.

The constraining action of the ecological functional units is analogous to the constraining action of enzymes, cellular membranes or organs in biological organisms. Like them, ecological units constrain the transformations of nutrients, which would otherwise occur only at a very slow rate or not at all. Both organisms and ecological systems – we claim – realise organisational closure. Yet, the crucial difference is that ecological self-maintaining networks are not, like organisms, an *autonomous* organisation because, unlike an autonomous system, an ecosystem as a whole does not exert higher-level interactive functional actions (in order to ensure its own persistence) on the external environment.

Let us explain this point. A set of interdependent constraints that realise closure necessarily implies a management of the thermodynamic flows that normatively serve the maintenance of that closure (and it is this management of the environment that we call "agency"). Since, in the case of ecosystems, the constitutive functional elements are *themselves* autonomous agents, the ecological organisation does not need to exert itself a higher-level of agency, for ensuring its own maintenance. At the metabolic level, closure requires agency as a set of constraints devoted to handling interactions with the external environment in accordance with intrinsic teleology and normativity; in turn, when closure is realised by an organisation of autonomous agents, the norm of the new (ecological) closed organisation could be simply the maintenance of adequate local environmental conditions for the different functional groups of autonomous agents.

In short, the hypothesis could be that the organisation of ecosystems realises closure without agency. Ecological networks are composed of autonomous (both unicellular and multicellular) agents, but as networks, they do not constitute autonomous entities because, crucially, they lack agency. Rather, what they do is acting in favour of better local environmental conditions, thus allowing more organismic diversity and providing greater support for their long-term continuity.

In this context an ecological function, in accordance with the characterisation developed in Chap. 3, is a specific effect of a constraint exerted by an ecological unit on the flow of matter and energy, within an ecological organisation. It is important to note here that we are assuming a very broad concept of "ecological unit" that we understand as the components of biodiversity. In specific terms and in the context of our definition of ecological function, something is an ecological unit if it is a biological entity whose activity is directly relevant for the maintenance of an ecosystem, actively participating in, at least, one constraining action within this same ecosystem (Nunes et al. 2014).

Unlike ecological interactions, strong inter-organismic relations could become much more complex and, as we shall see in Chap. 6, progressively "move up" towards new forms of autonomy. In the case of a multicellular organism, interactive relations between autonomous units become constitutive functions at the higher level. In turn, the higher level induces a deep-rooted transformation in its constitutive units. This is why, contrary to the case of ecological networks, the collective multicellular organisation possesses a stronger identity, possibly its own agency and, in some cases, higher level autonomy.

4.5.1 Toward Cognition

However important for the long-term sustainability of life, the phenomena dealt with in this chapter are manifestations of the basic form of agency. The most complex forms of agency, as we shall see in Chap. 7, appeared during the evolution of movement-based agency. Behavioural agency is the only form of adaptive agency that underwent an *open* process of complexification. Compared with movement-based agency, other forms of adaptive agency seem to "exhaust" their capacity for complexity growth.

Actually, the phenomenon of cognition (in our terms: cognitive agency) appeared as an evolution of behavioural agency. It is probably for this reason that behavioural agency (or at least behavioural agency when it has attained certain degrees of complexity) is much more widely and uncontroversially acknowledged as the most typical form of agency.

5
Evolution: The Historical Dimension of Autonomy

The mainstream view in contemporary philosophy of biology has largely considered the theory of evolution to be the main (or even the only) theory in the biological domain.[1] Accordingly, a large part of the philosophical debate has addressed issues such as "the structure of natural selection", the "units of selection", and the "concept of adaptation".

In evolutionary thinking, – and particularly in the framework of the Modern Synthesis –biological phenomena tend to be conceived as inherently historical in the sense that emphasis is placed on aetiological explanations – to employ once again Salmon's distinction introduced at the beginning of Chap. 2 – at the expense of constitutive ones.[2] As Dobzhansky's (1973) famous dictum claims, "nothing in biology makes sense except in the light of evolution". Here, we will examine three characteristic aspects of the mainstream view.

First, evolutionary approaches favour the idea that biological organisms essentially consist of "clusters of adaptations" and, as such, can be adequately explained by appealing to those evolutionary processes that have fixed such adaptations. What makes biological systems different from any other class of natural system is the fact that they are the result of evolution by natural selection (and other evolutionary processes), which explains novelty, diversification, and adaptations. All other features are superfluous to understanding what biological systems are

This chapter elaborates on ideas previously presented in Moreno (2007), Ruiz-Mirazo et al. (2008) and Moreno & Ruiz-Mirazo (2009).

[1] See for instance Sterelny and Griffiths (1999) and Sober (2006) for relevant overviews.

[2] We understand Salmon's distinction as mapping onto Mayr's one (more commonly evoked in the philosophy of biology) between "ultimate" and "proximate" causes (Mayr 1961). Accordingly, aetiological explanations would appeal to ultimate causes, while constitutive explanations to proximate ones.

A. Moreno, M. Mossio, *Biological Autonomy*, History, Philosophy and Theory of the Life Sciences 12, DOI 10.1007/978-94-017-9837-2_5

because they are not specific to the biological domain; hence, it is only evolutionary explanations that are relevant in accounting for not only the *genealogy*, but also for the very *logic* of biological systems.

Second, the prominent role of the theory of evolution affects the theoretical characterisation of what is to be taken as a "biological unit". At first glance, any entity upon which natural selection acts could be taken as a biological unit insofar as it exhibits variation, heredity, and differential fitness (Lewontin 1970). Consequently, a number of entities located at different levels of description, such as genes, organelles, cells, organisms, groups, communities, and ecosystems, could, in some cases, meet these requirements. Furthermore, the debate on the units of selection has been greatly influenced by Dawkins' defence of the gene-centred view of evolution, according to which the fundamental units of selection are genes (Dawkins 1976). Of course, there is widespread debate in the philosophy of biology on how units of selection should be understood, and we will not attempt to provide an overview here (see for instance Hull 1988 and Gould 2002). What is important for our purposes is the fact that, in this context, biological organisms do not possess a special theoretical status when compared to a number of other entities. Since all that matters for understanding biological phenomena are evolutionary processes, then not only are organisms essentially (clusters of) adaptations but, furthermore, they are not the only nor even the most relevant ones.

Third, biological organisation plays no role in shaping evolutionary processes. Evolution is fundamentally explained by natural selection exerted on a population of entities exhibiting variation, heredity, and differential fitness. In this view, biological organisation is the outcome of evolutionary processes, and is by no means one of its causes or conditions. Here again the debate on the structure of evolutionary theory is rich, and many authors have argued that other factors and processes should be integrated in order to extend the Modern Synthesis (Pigliucci and Muller 2010), and adequately account for evolutionary mechanisms processes in general – such as, for instance, genetic drift (Kimura 1968) and developmental constraints (Gould and Lewontin 1979; Maynard Smith et al. 1985) – and adaptations in particular (e.g. phenotypic plasticity, see Price et al. 2003). Yet in this literature, organisation as such maintains (at best) the status of *explanandum* of evolutionary processes, not that of *explanans*.

Again, the above characterisation is by no means an attempt at a detailed description of the rich debate currently being held within the mainstream evolutionary view thinking in the philosophy of biology. However, we think it appropriate to highlight some of its central tenets, to which most in the community subscribe, and on the basis of which we can contrast the mainstream and the autonomous perspectives.

How does the autonomous perspective conceive the historical-evolutionary dimension of biological systems? Let us return to the three features described above.

First, the beginning half of this book was aimed precisely at showing that biological organisation possesses several distinctive features that do not require an appeal to evolutionary processes (to ultimate causes, in Mayr's terms) in order to be described and understood. Indeed, the characterisation of biological organisms as autonomous systems points to the specificity of biological systems in their *current* organisation (in proximate causes, in Mayr's terms), regardless of the fact that they are also the product of natural selection. From the autonomous perspective, Dobzhansky's dictum would read: "Many things in biology make sense regardless of evolution"; the logic of the living is not reducible to its genealogy. In this respect, as we have pointed out in the introduction of this book, we agree with Varela and Rosen's remark, according to which answering the fundamental question "Why is an organism alive?" by appealing exclusively to evolution would amount to saying that "an organism is alive because its ancestors were alive". Biological systems are certainly the outcome of evolutionary processes, but this does not seem to be the entire story (see Introduction).

Second, the emphasis on biological organisation confers a privileged status on those entities that exhibit the features described in the previous chapters of this book, i.e. closure, regulation, and agency. In particular, among the whole set of entities belonging to the biological domain, organisms are the biological units *par excellence,* to the extent that they possess the relevant complexity required for realising autonomy (Ruiz-Mirazo et al. 2000). Of course, as we discussed in Chap. 4, while the characterisation of autonomous systems provided so far applies straightforwardly to unicellular organisms, it remains to be demonstrated how it also applies to multicellular organisms. This does not come without difficulties, and we will discuss the issue in more detail in Chap. 6. Nevertheless, we maintain that the autonomous perspective provides an alternative strategy for identifying biological units in theoretical terms: units of autonomy instead of units of selection.

Third, organisation is a *condition*, and not only an outcome, of evolutionary processes. As we will argue in this chapter, a sound explanation of the evolution of biological organisms by natural selection requires that we take their organisation into account for several reasons. One is that the principles of biological organisation constrain selective processes, which cannot lead to results incompatible with the conditions of maintenance of the organisation (and therefore of closure, regulation, agency, etc.): organisation reduces the set of possible outcomes of selection. At the same time, evolution towards biological complexity requires that selective forces be exerted on systems possessing at least minimal forms of organisation; otherwise, as increasing scientific evidence suggests, it would not lead to relevant results: organisation steers selection towards biologically relevant outcomes. Lastly, in realising closure, biological organisation can generate variations not only through random mutation, but also through its inherent activity and the interactions between its constituents: organisation itself generates novelty.

In broad terms, then, what is the relationship between evolution and autonomy, as conceived from the autonomous perspective? What role does history play? The general picture is that the evolution of biological systems stems from the mutual interplay between organisation and selection: in a word, organisation channels selective processes and selection drives organisation towards an increase in complexity.

It is important to stress that the autonomous perspective advocates the integration, rather than the opposition or tension, between the organisational and historical dimensions of biological systems. Although it places strong emphasis on the constitutive and interactive dimensions of biological systems, it does not follow that history is irrelevant: biological systems are *also* historical systems. While evolution requires (minimal) organisation, the realisation of full-fledged autonomous systems is the result of an historical process, through which changes and variations can be preserved and accumulated, enabling a progressive increase of complexity. The complexity of biological autonomy requires evolutionary processes.[3]

As a matter of fact, the historical dimension of autonomy allows a clear-cut conceptual distinction to be made between autonomous systems and self-organised systems. As discussed in Chap. 1, we hold that dissipative structures share the capacity to realise a (very minimal) form of self-maintenance with biological systems. Yet the analogies stop here. Dissipative structures possess a low internal complexity, which is precisely what enables them to *spontaneously self-organise* when adequate boundary conditions are met. Self-organising systems are systems that are simple enough to appear spontaneously. In contrast, autonomous systems are not spontaneous and cannot *self*-organise. Their distinctive functional ("organised", in Simon's 1969 terms) complexity goes far beyond that realised by dissipative structures, and cannot emerge "from scratch": each autonomous system is generated by another autonomous system, endowed with a sufficient degree of functional complexity and capacities. This is our rephrasing of Virchow's (1858/1978) motto "omnis cellula e cellula".

This is not to say that autonomy has nothing to do with self-organisation. Indeed, one of the challenges facing the autonomous perspective is to provide an explanation of how at least minimal forms of biological organisation emerged in the first place. Although autonomy requires history, we have suggested that relevant evolutionary processes require organisation; as we will see, the initial appearance of organisation (and specifically a closed organisation, in the form of a minimal self-maintaining chemical network) requires the interplay of both self-organising and self-assembling[4] processes. Still, those investigations into the preliminary stages of the origin of life concern the emergence of biological systems, not their intrinsic nature: autonomy may possibly proceed from (among other things) self-organisation, but it is not self-organisation itself.

It should be clear by now that the autonomous perspective, by appealing to the historical dimension of autonomy, advocates a framework that is complementary to the evolutionary one. Nonetheless, we would like to emphasise that one fundamental difference remains, as the theoretical status of history is not the same: although history is required to account for the emergence of autonomy, it does not

---

[3]Recently, Rosslenbroich has developed an original account of the relations between autonomy and evolution (Rosslenbroich 2014). Although largely complementary, his account is quite different from ours in that it mainly focuses on the evolution of the physiological changes leading to what he calls an increasing "independence of the environment". Accordingly, we leave a detailed analysis of his proposal for a future work.

[4]Actually, the concept of self-assembling mainly refers to the spontaneous formation of supramolecular (rather than purely molecular) structures in equilibrium or near equilibrium conditions.

define autonomy. Whereas the autonomous perspective refers to constitutive and interactive dimensions to answer the question "what is autonomy?", it appeals to the historical dimension to answer the question "how does autonomy emerge?". By this account, we do not need history in order to understand what biological systems are, we need history to understand where they come from: two related, yet distinct, questions.

The organisation of the chapter is as follows. First (Sects. 5.1, 5.2, and 5.3) we shall analyse the conditions required for the spontaneous appearance of a minimal form of compartmentalised closed systems (protocells) capable of harbouring at least a minimal functional differentiation, and discuss why this is not a consequence of, but rather a requirement for, natural selection. Next (Sect. 5.4), we will show how natural selection may have begun to operate even in the absence of a genetic hereditary mechanism once protocells have appeared. The evolution of protocells can, in turn, lead to a further increase in complexity and to the appearance of closed systems endowed with sequentially dependent components that could act as both hereditary templates and catalysts (Sects. 5.5 and 5.6). We will argue (Sect. 5.7) that this step still faced an evolutionary bottleneck, which was overcome with the emergence of specialised genetic constraints; that stage, we will submit, could correspond to the realisation of autonomy, which accords with the appearance of open-ended Darwinian evolution (Sects. 5.8 and 5.9), ensuring long-term sustainability of biological phenomena.

5.1 A Preliminary Look Into the Origins of Darwinian Evolution

The complexity of biological organisation – organised complexity[5] – goes far beyond that of any other natural system. Consequently, as stated earlier, it does not constitute a spontaneous phenomenon; rather, it is the result of accumulative processes that, starting from relatively simple systems, have produced a progressive increase in complexity.

In particular, current scientific knowledge conceives the origins of life as the result of the complexification of chemical systems that, in spite of their apparent fragility and instability, have been able to develop mechanisms for preserving functional innovations.[6] Understanding the historical dimension of autonomy implies,

[5] As stated in Chap. 3, Sect. 3.2.2, organised complexity is *functional* complexity.

[6] Human beings are used to building, maintaining, and managing complex structures and organisations. This could lead us to think that the generation of complexity is not a big issue. Yet, whenever we try to make complexity develop in a scenario in which there is no human presence, nor any possible intervention of other living organisms, things become much less easy. This experience coincides with what happens in the natural world, where (with the exception of life) systems show no great organised complexity: self-organising phenomena, for instance, create some self-maintaining dynamic patterns but are unable to increase this minimal complexity, whereas

then, an adequate account of these very mechanisms, which govern the generation and preservation of increasingly complex innovations. This is what the evolution of autonomy is about.

In the received view, selection gives rise to increasing complexity when operating on a population exhibiting variation, heredity, and differential fitness (Lewontin 1970). This is the conception that, for instance, lies behind the so-called "replication first" view in the origin of Life. According to this view, an evolutionary-selective path leading to the emergence of (proto-)biological systems can be generated in a situation in which there is competition between a set of self-replicating molecules capable of variation, and transmission of these variations through replication.

Are these conditions sufficient for the emergence of biologically relevant complexity? As we will argue in the following section, there seem to be good reasons for thinking that they are not. The central point is that selection cannot drive natural systems towards an increase in complexity unless these systems *already* possess a minimal degree of *organised* complexity. Systems or entities below the minimum threshold may indeed evolve by selection, but the resulting evolutionary path would not generate relevant complexity. (Low) organised complexity begets (high) organised complexity.

The autonomous perspective should therefore provide an account of the minimal conditions required for natural selection to drive biological systems towards an increase in their complexity. The objectives of the account would be twofold.

First, it should describe what kind of evolutionary processes, other than Darwinian selection, have brought about that minimal form of organisation. Darwin formulated his theory in the framework of full-fledged organisms, endowed with reliable genetic mechanisms of inheritance and in a context in which variation implies not only genetic change, but also (causally connected but clearly differentiated), phenotypic and functional variation. However, when we focus on the origin of life and therefore on much simpler systems, these requirements cannot be adequately met. Consequently, if natural selection cannot operate in a relevant way unless minimal organised complexity is already there, this means that, as several authors have pointed out (Fox Keller 2007; Godfrey-Smith 2009), there was a time when evolution was driven by other mechanisms able to generate that minimal complexity.

Second, the account should characterise the properties that biological organisation must possess in order to be able to increase its complexity through evolutionary processes. As we will see, this implies a sufficiently wide phenotypic domain upon which selection may act, as well as a reliable mechanism for the transmission of specificities (Moreno 2007; Ruiz-Mirazo et al. 2008; Moreno and Ruiz-Mirazo 2009). In the following sections, we will describe the transitions between increasingly complex forms of organisation – each endowed with some specific capacity for persistence and further evolution – right up to the emergence of those systems able to undergo evolution by natural selection as we know it.

certain assembling processes (like growing crystals) can generate and maintain a certain degree of structural complexity, but lack any form of functionality. Indeed, biological systems (and derivatively, human organisations such as social systems) constitute the only type of system we know of that can generate and increase both structural and functional complexity indefinitely.

The background assumption, again, is the inherent interplay between organisation and selection: organisation is not just the outcome of selective processes, but also a condition for selection to drive evolution towards the relevant degree and kind of complexity.

5.2 Replicative Molecules Versus Self-Maintaining Organisations

In the current debate about the origin of life there are two competing views: "metabolism-first" and "replication-first" (Pereto 2005; Anet 2004). The former holds that the beginning of biogenesis should be based on chemical self-maintaining networks (driven towards higher levels of complexity by principles of self-organisation), whereas the latter defends that Life began with the appearance of self-replicating structures (i.e. molecules), driven towards higher levels of complexity by natural selection.

The second view, as previously mentioned, relies on the idea that selection is more fundamental than organisation in the evolutionary emergence of biological systems. Typically, many scientists advocating this view assume that life started with a self-replicating molecule, the first "gene" or "replicator" that, when it appeared in adequate environmental conditions, would have rapidly generated a whole population of replicators, leading to a process of evolution by selection. Although a replicator is any entity that produces copies of itself, what these authors have in mind are in fact *modular templates*[7] (Maynard Smith and Szathmary 1995); namely, relatively complex oligomers that, by their structure, are able to catalyse their own copies (see also footnote 17). Since the specific order (and number) of the building blocks ("modules") of these self-replicating oligomers is not directly involved in their template capacity, sequential changes can occur during their replication. In turn, these changes are somehow hereditarily transmitted, thus leading to populations in which individuals differ with respect to the sequence of building blocks. It is then assumed that these differences may confer competitive advantages or disadvantages on these individuals.

The relevant question here is whether a longer and therefore more complex replicating entity would have a selective advantage. Contrary to what might be expected, the answer is likely 'no'. Let us consider why.

[7]Even very simple template replicators may show "hereditary" variations. Think, for instance, of the case of a self-replicating crystal, which by chance incorporates a screw-dislocation. Since this dislocation speeds up the binding of ions, it preserves its screw structure as the crystal grows. But in order to display an evolutionary process, advocates of the "replication first" hypothesis require the presence of modular self-replicating templates, namely, replicators possessing sequences of different building blocks, whose hereditary modifications will be considered the key element for displaying an evolutionary process (see Sect. 5.5 below).

Any evolutionary scenario based only on a population of replicating molecules will have serious difficulties in increasing or even maintaining its complexity. The problem is not just that sooner rather than later replicators would need compartments to avoid the problem of parasites, nor that there is a critical length of nucleotide chains above which it is no longer possible to carry out a reliable replication process (Eigen 1971). More significantly, it is hard to conceive of the very appearance, maintenance, and evolutionary development of populations of self-replicating molecules in the absence of a rudimentary metabolic organisation producing an open space of functional variations, such that changes in their molecular sequences have a wide enough range of dynamic-operational effects within the system. The point is that if molecules do not belong to an organisation (i.e., if they are not the components of a more encompassing self-maintaining entity), the only way they have to raise their fitness is by improving their individual replication rate or their resistance to hydrolysis. And alone, this selective pressure does not seem to drive towards an increase in complexity, as suggested, for example, by Spiegelman's experiments (see also Chap. 4).

In 1967, Spiegelman conducted a set of experiments during which the RNA from a simple virus (Qβ) was inserted into a solution that contained the enzyme RNA replicase from the Qβ virus, some free nucleotides, and some salts (Mills et al. 1967). In this environment, the RNA started to replicate. After a while, Spiegelman took some RNA and moved it to another tube containing a fresh solution. This process was subsequently repeated. Shorter RNA chains were able to replicate faster, so the RNA became shorter and shorter. After 74 generations, the original strand with 4,500 nucleotide bases ended up as a dwarf genome, which was called the "Spiegelman's Monster", with only 218 bases. Such a short RNA had been able to replicate very quickly in these circumstances (Kacian et al. 1972).[8] This experiment suggests that in the absence of some organisation providing a sufficiently rich phenotypic domain, selective forces cannot perform beyond a minimal space of action. Thus, a scenario that provides enough phenotypic variety to be selected for seems to be required in the first place (Wicken 1987).

Advocates of the "replication-first" approach may object that given adequate environmental variety, selection alone may favour an increase in the structural complexity of the replicators. For example, Pross (2003) has argued that:

> Of course, the emergence of more complex replicators would not be kinetically sustainable if the added complexity were unable to provide some kinetic advantage – complexity must provide some existential advantage. It now seems clear that the kinetic advantage that longer sequences could provide would not have stemmed from any inherently greater replicating ability associated with the longer sequences (Spiegelman's experiment demonstrated that) but, rather, *through a variety of catalytic effects that some particular sequences might have afforded* (our italics, 401).

[8]Thirty years later, Oehlenschläger and Eigen (1997) showed that the Spiegelman monster eventually becomes even shorter, containing only 48 or 54 nucleotides, which are simply the binding sites for the enzyme RNA replicase.

Pross' argument, however, implicitly admits that the functional domain required by selective mechanisms to drive systems towards higher degrees of complexity is linked to what he calls a "variety of catalytic effects". To us, this points straightforwardly to the idea of a catalytic network or, as discussed at length in Chap. 1, to organisational closure: namely, in order to achieve at least a minimal form of functional diversity, a type of organisation we have called "organisational closure" is required. Accordingly, Pross is referring to a rather different scenario, in which populations of molecules, instead of *competing* for faster replication, have diverse catalytic effects on each other as a way of *coordinating* the particular locations, times, and speeds at which their chemical transformations occur. This implies a gathering together of certain reactions (i.e., embedding the synthesis processes of new structures) and a degradation of others in a self-maintaining organisation. Thus, in the end, many advocates of the primacy of replication and selection seem required to accept the almost immediate inclusion of organisational features in their framework if it is to account for any increase in complexity. Hence, it seems that any form of relevant evolution in the context of the origin of life requires a self-maintaining organisation as a starting point, harbouring a minimal functional differentiation (Moreno and Ruiz-Mirazo 2009).

With this preliminary conclusion in hand, the subsequent question is what kind of specific organisation is a relevant candidate for bootstrapping a Darwinian evolutionary process. The fundamental problem consists of understanding how a class of systems can be organised so that given certain specific but probable conditions,[9] it fulfils the three following requirements (Moreno 2007):

1. It is, in principle, simple enough to emerge spontaneously from a set of material aggregates;
2. It possesses the capacity to increase its functional variety;
3. It is able to preserve functional innovations.

As for the first requirement, we need to conceive a set of plausible boundary conditions which would enable the spontaneous appearance of chemical systems endowed with, in at least a very simple way, a self-maintaining closed organisation, although such systems were probably preceded by many other systems whose maintenance was essentially dependent on boundary conditions being much more complex than themselves.

As for the second requirement, the system must be able to increase the number of functions. Not only must the organisation be able to spontaneously generate new internal differences but, moreover, some of these differences should later play a new functional role. In turn, new functions engender new forms of organisation (say, a self-enclosing autocatalytic network), which might be preserved if they allow for more stable maintenance. In this respect, self-maintaining organisations are of particular importance due to their *inherent capacity to increase their complexity*.

[9]In terms of what the physical and chemical evolution of the universe can create in certain places during reasonable periods of time.

Indeed, since the different constraints exist because they make a contribution to the maintenance of the whole organisation of the system, variations that do not destroy the organisation could in turn act as new constraints, generating more sophisticated and accurate functions provided the whole set of constraints subject to closure is able to maintain itself (Mossio and Moreno 2010).[10]

This leads us to the third requirement. Along with the capacity to generate functional variety, a sustainable process of increase in complexity requires methods of preservation. The preservation of functional complexity requires a mechanism that ensures its maintenance beyond the lifespan of individual systems. In turn, this requires reproduction processes coupled with some forms of heredity. Thus, organisational closure has to be complemented by self-reproduction (for example, by growth and fission) and this reproduction has to ensure at least a minimal degree of specificity in the cross-generation transmission. It is worth emphasising that in these early steps, there is no need for reliable heredity, since some form of statistical transmission of specificity may suffice.

In the following section we deal with the first two requirements: how has closure first appeared; and how could it have been able to increase its complexity?

5.3 At the Origins of Organisation: The Emergence of Protocells

Some minimal form of organisational closure is therefore a requirement for starting any relevant evolutionary process because, as we have argued above, selective forces require functional variety to operate and drive the increase of organised complexity. What kind of processes could lead primitive chemical systems towards a minimal form of organisational closure?

To date, no comprehensive account of the evolutionary emergence of minimal closure exists. These are issues that, arguably, depend fundamentally on empirical research. Yet, it is likely that a wide variety of chemical systems appeared in specific places on the planet during the period of chemical prebiotic evolution that took place when the Earth cooled down. Current scientific knowledge supports the hypothesis that some local environments of our primitive Earth (e.g., hydrothermal vents) may offer adequate and relatively stable conditions for relevant chemical evolution processes to occur: a constant flow of energy and micro-porous surfaces, for instance, would have favoured the appearance of far-from-equilibrium chemical cycles leading to the formation of relatively complex organic compounds (Martin and Russell 2003) (Fig. 5.1).

Moreover, as Fox Keller (2007, 2010) has argued, in the early steps of prebiotic evolution, before the appearance of Darwinian Selection, simple self-maintaining

[10]To see why organisational closure is a crucial requirement for the increase of functional complexity, compare the situation above with that of a minimal dissipative structure, such as the flame. In this case, a variation of some component (the various material structures involved in the flow) does not affect the behaviour of the other, because it does not exert any specific causal

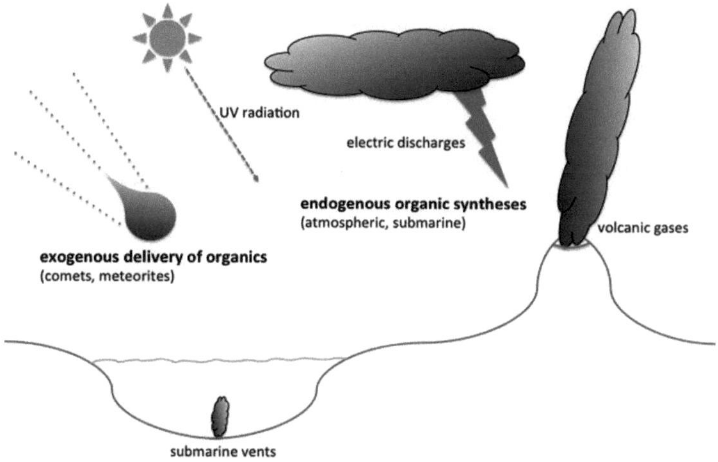

Fig. 5.1 Schematic figure of the geochemical conditions for the appearance of the early self-maintaining systems in our planet (*Credits: Juli Peretó*)

systems might have evolved towards greater complexity through what she calls a "selective process for stability and persistence", i.e. a selective process that favours more stable systems. Indeed, these "precursor reaction networks" might potentially realise a (very) minimal form of closure in the form of collective catalysis. Although very important for explaining the availability of certain chemical species in prebiotic times, these networks do not seem to constitute the relevant starting point for the evolution of complexity, insofar as they would be too directly dependent on (or immediately exposed to) environmental conditions (thermodynamic flows, diffusion forces, etc.). For example, they could have occurred on mineral surfaces, or through association with externally formed vesicles. But, of course, these external constraints would not be subject to closure, and therefore would not be generated and maintained by the network. In a word, the level of complexity that these networks can attain still appears to be severely limited, and unable to meet with the second requirement formulated above.

In turn, the evolution towards higher degrees of functional complexity may indeed have been bootstrapped by the phenomenon of *compartmentalisation,* which occurs when a self-maintaining network makes a contribution to the production and/or maintenance of the vesicle that encapsulates it, such that the latter can be said to be included in closure. Through compartmentalisation, externally provided constraints might have been progressively "recruited" and become a functional part of the internal organisation of the system. As several authors have pointed out (Deamer 1997; Luisi 2006; Luisi et al. 2006; Mansy et al. 2008), self-assembling boundaries, like vesicles, are quite plausible supra-molecular structures on primitive

contribution to the maintenance of the whole. Because of this, the flame will keep behaving in the same way in spite of various possible modifications of its components.

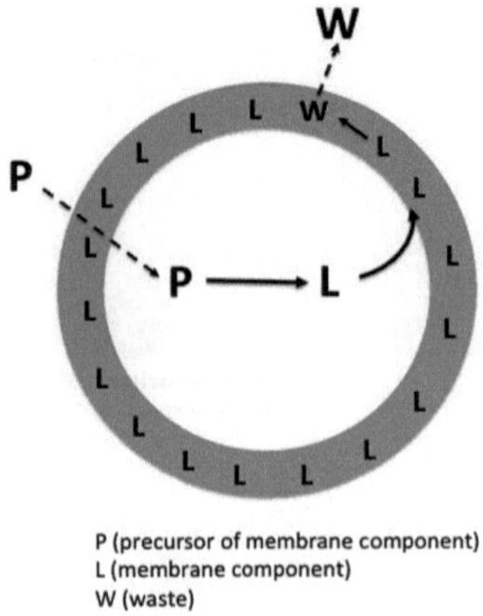

Fig. 5.2 An abstract scheme of a protocell, in which a reaction inside produces the building blocks of the compartment (*Credits: Juli Peretó*)

Earth; it is very likely that these compartments were associated with reaction networks. In the literature, these systems are referred to as *protocells*[11] (Rasmussen et al. 2008) (Fig. 5.2).

The importance of compartmentalisation can be assessed against the fact that all biochemical reactions, as we know them, do occur in compartments and distributed domains that guarantee the specific internal conditions required for metabolisms to run. It seems that increasingly complex reaction networks could only develop within compartments under suitable physicochemical constraints that enabled, for instance, high enough local concentrations, as well as a control of the flow of matter and energy through the system (selective in-and-out permeation of certain metabolites, energy transduction mechanisms, etc.). Indeed, a complementary relationship exists between the evolution of the structure of the membrane and its internal network (Morowitz 1992), since the components of the former selectively modulate its permeability and this in turn allows more complex networks inside. The reverse is likewise true, i.e. once the internal network reaches a certain threshold of complexity, it can play a role in the production, maintenance, and even the reproduction of the membrane. We therefore hypothesise that compartmentalisation constitutes a crucial requirement for the emergence of a form of minimal organisational closure that, in turn, can be a relevant step for the further increase of organised complexity.

It is worth emphasising that the increase of *organised* complexity relies on the capacity to generate increasingly higher degrees of *structural* complexity. Indeed, as we will see later on, the emergence of more precise and diverse functions in chemical self-maintaining networks depends on the development of stereospecificity

[11] A protocell is any experimental or theoretical model that involves a self-assembling compartment linked to chemical processes taking place around or within it. The model is aimed at explaining how more complex biological cells or alternative forms of cellular organisation may come about (Ruiz-Mirazo 2011). Here, we use the concept of protocell in a slightly more specific sense, as a compartmentalised closed system showing some lifelike properties, such as growth, autocatalytic activities, or reproduction (Rasmussen et al. 2008).

(a particular spatial configuration of the components of a large molecule), which, in turn, requires a substantial increase in size and structural variety. Although the formation of relatively complex structures could have been driven by geological or other types of abiotic processes in different environments, an accumulative production of complex structures leading to the appearance of relevant macromolecular structures requires precisely organised systems, as described above. The reason is that within organised systems, increasingly complex structures could be recruited to perform functional activities and, because of closure, maintained through the operation of repair and reproduction by the organisation itself.

The consequence of this mutual interplay between spontaneous phenomena of both self- organisation and self-assembly is the emergence and maintenance of new, larger structures, made up of many simpler functional components. As Deamer (2008) puts it:

> The result of this process would have been that vast numbers of microscopic assemblies of molecules appeared wherever organic compounds became concentrated at the interface between the atmosphere, water and mineral surfaces. In one scenario (. . .) these assemblies took on a cell-like form (. . .) each cell-like assembly had a different composition from the next. Most were inert, but a few might have contained a particular mixture of components that could be driven towards further complexity by capturing energy and small nutrient molecules from the environment (. . .) As the nutrient molecules were transported into the internal compartment, they became linked together into long chains in an energy consuming process.

The central point is that the increase in structural complexity leading to new functions in a system would primarily result from compositional arrangement processes of the "primitive" functional parts, namely, the differentiated constraints involved in the organisational closure of the system. Through several processes of self-assembly, the resulting structural complexity would be capable of triggering new structural changes, thus opening up new possibilities for subsequent composition in an open process of recombination.

As a result of the interplay of both self-organising and self-assembling phenomena, a variety of compartmentalised closed systems with different degrees of complexity and robustness might have spontaneously appeared in certain local plausible conditions. These systems could harbour the capacity to generate more complex structural components that, in turn, might have enabled the increase of functional complexity. Once this kind of protocell has emerged, some of which may have been capable of self-reproduction,[12] a process of evolution would have been initiated. Let us turn to this in the next section.

[12]Compartmentalisation could actually have induced self-reproduction. It might have been the case that, in some circumstances, chemical self-maintaining network developed within some, pre-existing available empty vesicles, so that the resulting system might have enlarged the vesicle until it slopped some of its chemical content over into a neighbouring vesicle; in turn, the chemicals could have slowly re-formed the original self-maintaining network (Hooker, personal communication). In this chapter, we do not discuss the specific conditions in which the reproduction of protocells might have emerged. We simply suppose that this step has been made at some point.

5.4 The Origins of Natural Selection

As soon as reproduction is included in the scenario of the evolution of protocells harbouring a minimal form of functional diversity, we shift to a populational perspective.

Assuming that there is also a certain form of inheritance (as we will discuss next), the appearance of populations of protocells could have led to a primitive form of evolution by natural selection; this in turn could have favoured those systems whose functional integration happens to be more efficient, while eliminating others. There are a variety of combination in which the functional components of these systems may contribute to their maintenance and reproductive success. This variety ensures a wide enough phenotypic space for selection to actually operate as an evolutionary mechanism without running into "dead ends" or bottlenecks of too low complexity.

Budin and Szostak (2011) have recently depicted an intriguing scenario of the appearance of natural selection; they have specifically tried to identify mechanisms of competition amongst protocells in relation to the structure of their membrane. Considering the low complexity of membranes in the early stages of prebiotic evolution, they enquire into the selective advantage that may have driven the evolution from self-assembled, simple, single-chain lipid membranes to phospholipid membranes. They argue that, according to the results of their research:

> phospholipid-driven competition could have led early protocells into an evolutionary arms race leading to steadily increasing diacyl lipid (e.g. phospholipid) content in their membranes (ibid.: 5252).

Protocells could have started to evolve membrane transporters along with protometabolic networks for synthesising their own building blocks and may have begun exploring new environmental niches compatible with compounds that otherwise decayed rapidly in fatty acid membranes. They conclude that the transition from highly permeable vesicles to less permeable and more stable protocells was driven by a primitive kind of selection, resulting in the evolution of the functional domain of the protocells. In this respect, the appearance of phospholipid membranes was a crucial step towards the internal control of the conditions that favour the further increase of organisational complexity.[13]

Reproduction and competition, however, are not, in themselves, enough. Evolution by natural selection would also require a minimal form of *inheritance*. How could a mechanism of inheritance have emerged before the appearance of more sophisticated mechanisms that include genetic components? A likely example of primitive inheritance may be, as Segré and co-workers have suggested (Segré and Lancet 2000; Segré et al. 2001), the so-called "compositional genomes": namely, compositionally biased catalytic networks, devoid of sequence-based biopolymers,

[13]For a detailed discussion of how a minimal form of functional diversity could arise in this scenario, see Arnellos and Moreno (2012).

capable of transferring their compositional specificity through reproduction. As these authors have argued, specific organisational features of protocells could be inherited through generations because different types and quantities of molecules could be *statistically* transmitted.

The hypothesis of compositional genomes has received several criticisms that focus primarily on the limitations of the "lipidic world" on which it is based. In particular, there does not appear to be enough catalytic variety permitted in a scenario where protocells are based on lipids. More recently, however, Vasas et al. (2012) have considered a much richer polymeric world (a world of polypeptides, as opposed to the lipidic world of Segré and Lancet) and have shown that in this scenario some selective processes could take place. As the authors themselves write:

> Our work shows that autocatalytic sets as first devised by Dyson and Kauffman are theoretically possible despite previous criticisms and, perhaps more interesting, that chemical evolution in these systems can lead to the appearance of viable autocatalytic cores, thus opening the possibility for evolution by natural selection. (...) After all, the pre-template Darwinian dynamics of rare core production and selection described here (...) is the only viable proposal so far for how autocatalytic reaction networks could accumulate adaptations.

Accordingly, populations of reproducing protocells may possibly realise a pre-genetic mechanism of statistical inheritance. This in turn creates the potential for a form of evolution by natural selection in which organised complexity can be enhanced and preserved. Yet, in these conditions the complexification driven by natural selection still faces certain limits. The degree of complexity such protocells can attain is limited, in particular because there is no mechanism to ensure the production and maintenance[14] of polymers (long monomer chains) with specific aperiodic (non-redundant) sequences, which would lead to the development of stereospecificity, inter-molecular recognition mechanisms, and catalytic efficiency (based on folding and the more elaborate chemistry of multiple weak bonds). These types of sequentially specific polymers are necessary for performing more specific catalytic tasks and, in particular, for enabling regulatory functions. In the absence of regulation mechanisms, protocells face a major bottleneck: the higher the molecular complexity in the system, the more difficult it becomes to ensure robust self-maintenance. As complexity grows, so does fragility (see also Chap. 1, Sect. 1.8). And since most functional and regulatory roles are linked to new, sequentially-dependent oligo/polymers, protocells cannot increase their complexity beyond a certain degree.

In summation, the scenario sketched so far accounts for the emergence of compartmentalised, organisationally closed systems (protocells) endowed with a certain degree of functional diversity and relatively complex components (probably oligomers made of short functional sequences of building blocks). From a biological perspective, these systems would still be extremely simple and lack most biological

[14] Szathmary (2006) has analysed the limitations of this form of pre-genetic mechanism of inheritance. We shall come back to this issue in next sections.

properties, even in comparison to those exhibited by current prokaryotic cells. Protocells would nevertheless be able to reproduce, possess some primitive forms of statistical inheritance mechanisms, and undergo evolution by natural selection. Yet their constitutive organisation cannot ensure that the structure of their functional components will remain unaltered for much longer than the lifespan of each individual system. At this stage, the evolution of protocells faces a bottleneck: as organised complexity increases, its preservation becomes more and more difficult. Therefore, only those systems that developed specific mechanisms to stabilise and retain their increasing organisational (and, hence, structural) complexity with a fairly high degree of reliability could begin to unfold new and higher degrees of complexity and, furthermore, establish the groundwork for ensuring their long-term maintenance.

Since the emergence of reliable heredity supposes the appearance of much more complex components, in the next section let us first discuss how structural complexity may have further increased and at the same time have been preserved. Although we shall discuss these two processes separately, it is highly likely that their respective evolution has been closely interwoven.[15]

5.5 Early Forms of Template-Based Evolution

Protocells, as discussed, could have evolved towards higher degrees of organised complexity, in particular involving the capacity to generate and maintain sequentially specific short molecules (oligomers).

To achieve further degrees of complexity, protocells required the invention of a new form of organisation that resolved two interrelated problems: on the one side, the production of a diverse set of highly efficient catalysts (first, able to perform specific functions; and second, regulatory functions), which would enhance metabolic performance and versatility; on the other side, the reliable preservation of the increasingly complex functional components within the organisation, providing robustness as well as hereditary stability.

As for the first problem, namely, the production of efficient catalysts, let us see how the continued increase in structural complexity is possible. Essentially, the starting point is the availability of relatively stable entities that are amenable to a variety of physical forms of assembly or aggregation, such that they can act as "building blocks" (through compositional processes of rearrangement) for the

[15] In fact, current scientific research into the origin of life increasingly supports a synthetic view in which the three key questions – the formation of a proto-metabolic organisation, the creation of a selectively permeable compartment for this organisation, and its reliable hereditary reproduction – appear deeply entangled and therefore influencing each other (Ruiz Mirazo et al. 2013).

construction of an open world of more complex functional variants.[16] As Fox Keller (2009) has pointed out:

> The formation of the covalent and non-covalent bonds that hold such molecular complexes together can also sometimes change the structure of the components with which the process started. In so doing, they can also induce changes in the rules of engagement, thereby creating the possibility for new interactions, new binding sites, new hooks. The new binding sites are not simply the consequence of the new proximities created by molecular binding, but more interestingly, of the changes that have been triggered in the ways in which the component parts can interact. They might be thought of as Brownian motors in evolutionary space, feeding on chance events to build ever more complex configurations (...) The phenomenon I am trying to describe rests on two basic facts: first, that many complex macromolecular structures are capable of stabilizing in a variety of distinctive shapes or forms, and second, that the binding of new molecules can trigger a shift from one conformation to another, thereby exposing new binding sites, and new possibilities for subsequent composition (*ibid* pp. 22–23).

This process opens up a new and rich domain of structural variety and, specifically, the capacity to generate more complex molecular aggregates, such as polymers that are constituted by a specific sequence of elementary building blocks (Srere 1984). The specific order of the building blocks (monomers) contributes to the determination of the shape of the polymer, which in turn determines its catalytic properties.

Once the specific (and highly unlikely) sequential patterns supporting catalytic capacities have been discovered, protocells must *fix* them in their organisation. If they did not, they could not be maintained and their potential advantages would be lost in a few generation steps. This leads us to the second issue, namely, the preservation of the structural and organisational complexity. The only way to retain new functional patterns of such high structural complexity seems to be through establishing some kind of "template" or "blueprint" copying mechanism, which ensures the replication of their particular sequences (either exactly or almost exactly). As Szathmary and Maynard Smith (1997) have pointed out, the preservation of long and complex polymers requires what they call a mechanism of "unlimited memory". The mechanism consists of linking the sequential structure of certain stable components (in particular, of modular templates[17]) to the more

[16]Actually, it seems that quite small molecules could act as building blocks. For example, as shown by Manrubia and Briones (2007), certain small molecules of RNA can play the role of modules in a stepwise model of ligation-based modular evolution: RNA hairpin modules could have displayed ligase activity, catalysing the assembly of larger, eventually functional RNA molecules. These ligation processes allow a fraction of the population to retain their previous modular structure, and thus structural and functional complexity can progressively increase.

[17]A molecule acts as a template if its structure acts as blueprint, enabling the formation of copies of said structure. *Modular* templates (Maynard Smith and Szathmary 1995; Szathmary 2000) consist of interchangeable discrete units, which build up a specific one-dimensional sequence, and whose global three-dimensional shape is such that it allows the recurrent copying (by a chemical complementarity mechanism, like base pairing) of complete, equivalent sequences. Although modular templates are considerably complex molecules, simple kinds of templates probably played an important role in previous evolutionary stages.

complex structural properties of a functional polymer. The templates play the role of reliable (and highly specific) constraints that, when replicated, enable a reliable form of reproduction of the complex polymer. Unlike simpler templates (like the ones present in the growth of a crystal), modular templates can enable the generation and maintenance of an indefinite amount of structural complexity.

It is now widely accepted that, before the introduction of the highly inert DNA, RNA that was much more catalytically active played the role of early modular template. This evolutionary stage corresponds to what is currently called "the RNA world".[18] The specificity of RNA is that it could carry out both template and catalytic tasks within the cell. Therefore, besides reliably constraining the replication of sequences, RNA could also directly convert specific sequences into catalytic functions.[19] In contrast to more simple protocells, in organisations endowed with RNA templates, increasingly complex catalytic functions are specified by the linear sequence of some components.

The appearance of individual closed organisations based on sequentially specific RNAs triggered, in our view, a key transition from "proto-metabolic" systems (protocells) to what certain authors consider as genuine metabolic ones. De Duve (2007), for example, uses the term "proto-metabolism" to refer to those chemical networks driven by catalysts that, whatever their nature, cannot have displayed the efficiency of sequentially specified enzymes or ribozymes. What would these prebiotic systems look like at this stage? Somewhat speculatively one may argue that, given the fact that RNA catalysts are (for different physicochemical reasons) much less capable of supporting catalytic functions, RNA-based protocells would still be much simpler than present-day prokaryotic cells. Since RNAs could hardly support regulatory mechanisms – in fact, even the simplest regulatory mechanisms that are known in prokaryotic cells are based on proteins – these systems would probably lack regulatory functions. As a consequence, their interactive functions would presumably not be adaptive: in sum, they would *not* fulfil the requirements for autonomy.

In turn, RNA-based protocells could be able to set complementary metabolic exchanges, such that it is sensible to suppose that their interactions might have generated primitive ecological networks.

[18]It is important to clarify that by "RNA world" we refer here not to "nude" self-replicating RNAs, but to closed organisations whose metabolism was catalysed by RNAs and whose reproduction was specified by RNA templates.

[19]RNA, however, cannot perform both functions in a very efficient way. We shall explain this point in the next section.

5.6 On the Nature of Template Constraints

So far, we have characterised protocells as closed systems whose organisation is constituted by a set of internally generated (and maintained) constitutive and interactive constraints. The appearance of sequentially specific polymers (like RNAs) playing both the role of templates and catalysts changes the situation. For if we ask where the specific *sequence* of these components comes from, the answer would point to a system that is more encompassing than the individual organism in itself. Although they are made up of building blocks just like any other component and are subject to organisational closure, the specific order of sequential components is ultimately a consequence of the evolutionary process driven by natural selection, which goes beyond the frontiers of individual organisms.

Let us develop this point. Given that the catalytic function of these modular components depends crucially on the specificity of their sequence, the preservation of this specific order becomes fundamentally important. The fact that these components are at the same time templates is crucial because, as a result of this property, they ensure the inter-generational transmission of this specific order: they are hereditary constraints, which can reliably preserve organisational changes from one system to another.[20]

Over longer periods of time (i.e. many generations), the sequential order of these hereditary constraints may undergo changes. All those changes that allow viable reproduction will lead to an exploration of the sequential space linked to a correlative selective retention of the organisational forms. This allows individual systems to recruit the results (i.e. selected patterns) of a slow process of natural selection, which encompasses these same individual systems both temporally and spatially. The evolutionary process in which the whole population and its environment are involved largely determines the changes affecting the template sequences. In this way, organisms endowed with modular templates can coherently and consistently link the individual dimension of their activity (related to their constitutive closure) to a progressively larger temporal and spatial dimension (related to their long-term maintenance and evolution as a whole population). Globally speaking, therefore, the template-based closed organisations integrate two temporal and spatial dimensions. In some sense, this articulation is enabled by the inherited sequential structure of these special functional components (i.e. RNAs) of each individual entity.

To avoid misunderstandings, it should be emphasised that the domain in which the sequences are shaped is not to be conceived as independent from the dynamics of each individual organisation, insofar as the very activity of templates does not make sense separate from the organisation as a whole; moreover, the fitness of some sequence depends of course on how the whole system works and interacts

[20]Of course, this is not to say that during reproduction, inheritance mechanisms concern these templates uniquely, rather, that their importance lies in the fact that (1) they can "localise" the hereditary changes, and (2) they ensure the structural specificity of the most complex functional polymers of the system.

with the environment.[21] Yet, our main point here is that the beginning of a process of evolution, including the transmission (both vertical and horizontal, as suggested by Woese[22]) of modular templates, inaugurated a stage in which individual (proto)organisms have a crucial dependence upon the long-term selection of functional hereditary components. In turn, this global process of selection will depend on the performance of the individual proto-organisms that it contributes to specify. With the appearance of reliable hereditary templates, the maintenance of the increasingly complex organisation of these proto-organisms will be inherently linked to the historical and collective web they are weaving.

5.7 The Emergence of Specialised Template Functions

Presumably, RNA-based protocells were the immediate precursors of present-day living organisms.[23] As discussed above, the organisation of these systems was likely based on a single type of polymers that supported both template and catalytic functions. Yet, this kind of organisation cannot lead to an unlimited evolutionary increase in complexity because it involves a trade-off between the realisation of catalytic and replicative functions. Indeed, the better suited a given type of polymer is for template tasks, the worse it is for exploring the catalytic space, and vice-versa,[24] which means that neither a full exploration of the sequential domain nor a full conversion of sequential variation into new functions are possible.

---

[21]The particular "performance" of a given metabolic organisation in a specific environment, and hence the capacity of this system to successfully reproduce, is dependent on the nature of the functional constraints that constitute this system. In this sense, selection operates on the organisation as a whole; but because (at least in certain cases) changes in hereditary records are linked to localised changes in functional constraints, selection could also be phenotypically specific.

[22]As Woese (2002) has pointed out, the beginning of cellular evolution was a collective process, where different cellular designs evolved simultaneously, systematically exchanging genetic material (what he calls "horizontal gene transfer"). So, this early (pre)Darwinian evolution would allow an exploration of different forms of organisation, until a "modern design" was reached.

[23]The current view of the origin of life postulates a stage of prebiotic systems based on a certain type of bi-functional polymers (like RNAs) capable of performing both template and catalytic functions, although in a much less suitable way than DNA and proteins. Hence, despite its evident limitation in the exploitation of both template and catalytic functions, this solution is organisationally much simpler (since it allows the direct conversion of a specific sequence into a specific catalytic task) and is therefore more likely to have occurred.

[24]This problem has a simple chemical interpretation. Template activity requires a stable, uniform morphology, suitable for linear copying (i.e., a monotonous spatial arrangement that favours low reactivity and is not altered by sequence changes); whereas catalytic diversity requires precisely the opposite: a very wide range of three-dimensional shapes (configuration of catalytic sites), which are highly sensitive to variations in the sequence (Moreno and Fernández 1990; Benner 1999).

Biological evolution has overcome this limit by introducing two different types of polymers, devoted respectively to replication and catalytic tasks. In this new kind of organisation, the former template-catalytic components (RNA) are replaced by two others: specialised templates, completely free of any catalytic task, which become tools for an unlimited memory (as we know them in present-day DNA); and specialised catalysts, better suited for translating sequential variations into three-dimensional diversity (as indeed occurs in present-day proteins).[25]

It must be stressed again that the differentiation between these two kinds of functions can only occur within a *common metabolism*[26] that is responsible for maintaining a constant link between the two and enables their complementary development within individual systems and over the course of generations (Ruiz-Mirazo et al. 2008). In particular, the duplication of DNA requires the action of a whole family of DNA-polymerases, and the transcription of DNA into mRNAs – and ultimately into proteins – requires the action of RNA-polymerases, tRNA-synthetases, and other proteins. Reciprocally, all these molecules depend on DNA, because they cannot be re-synthesised without the latter. Since templates and catalysts are not made of the same kind of monomers (or the same kind of chain bonds) an indirect, mediated connection becomes a requisite for ensuring effective interaction between the template and catalytic functions. The connection established by the metabolism as a whole corresponds to what is usually referred to as the "genetic code". From the autonomous perspective, therefore, the genetic code becomes *the expression of an organisation*, rather than a set of rules on which a local mechanism is supposed to operate.[27]

In the new form of organisation, which harbours the distinction between templates and catalysts, template constraints constitute a particular kind of function. As any other functional components, they are subject to closure and therefore intrinsically dependent on their causal connection with the whole organisation. Yet, with respect to other functions, they operate at a larger time scale, which means that they are relatively decoupled (rate-independent) and stable, in dynamic terms, with respect to the on-going metabolic chemical reactions (Pattee 1977).

Moreover, their causal action is quite peculiar. Specialised templates, as DNA, indirectly constrain metabolism by selecting specific sequences for the different amino acids building up proteins, which enables the synthesis of otherwise highly

[25]However, RNAs have not been erased by this new world of DNA-proteins, since they still play a crucial role in the complex relations between these two radically different polymers.

[26]Here, as explained below, we use in accordance with the definition given in Chap. 1, i.e. as a closed and *regulated* organisation.

[27]Using a linguistic terminology, Pattee (1982) has emphasised the fact that this relation is also subject to organisational closure. According to him, the genetic code should be understood through the idea of "Semantic Closure". Pattee considers that gene strings are self-interpreting symbols because their action (specific but arbitrary because it is mediated by the recognition of certain functional components) is the synthesis of those components (tRNAs and synthetases) that allow the causal action of the genes themselves. Thus, by contributing to the maintenance of the whole cellular organisation, genes in fact achieve their own interpretation.

improbable proteins. Actually, DNA is an extremely passive molecule.[28] It is only through a series of functional actions exerted by ribozymes and proteins (first converting DNA strings into RNA strings, and then, within the ribosomes, converting the RNA strings into amino acid strings and proteins) that the sequence of DNA functionally matters. In particular, if we consider the relevant time scale at which all this machinery can be taken as a kind of "black box", the DNA as a whole can be pertinently described as a constraint exerted on the synthesis of proteins: indeed, DNA participates in the harnessing of the specific distribution of the amino acids forming the building blocks of the proteins, while being conserved with respect to this overall process.

The highly stable nature of DNA, as well as its dynamic decoupling from the metabolic processes, enables us to view changes in DNA sequences as largely independent of the metabolic organisation itself. Ultimately, the decoupling of template functions from the metabolic dynamics is the expression of the inherent insertion of organisms, as autonomous systems, into a historical-collective dimension where the "slow" processes of creation and modification of evolutionary patterns take place, and where the mutual dependence between individual organisation and the eco-evolutionary dimension is better established. Each time a new template structure linked to the production of a new functional protein enters the organisation of a cell, providing this modification turns out to be viable and advantageous for that cell, a new causal link becomes stabilised. Thus, the template components, shaped through a collective and historical process, re-arrange material subsets of structures so that highly organised systems are generated and preserved. One important feature of this new kind of organisation is that the specification for the maintenance of the system is hierarchically organised: a significant part of these specifications are constrained by the sequences of the templates. This allows the robust maintenance of much more complex networks (which in turn will support more specifications in their connectivity), and self-sustained feedback between templates and metabolic networks, leading to further increases in complexity (Ruiz-Mirazo and Moreno 2006).

The differentiation between templates and catalysts, we submit, opens the way to the realisation of autonomy. As we have argued, the organisation endowed with proteins can potentially explore a huge sequence space of the modular templates and, possibly, of functional innovations. In particular, proteins can support sufficiently higher degrees of structural complexity, so as to provide not only a huge number of different and very specific functions in the system but also, and crucially, *regulatory* functions, as happens in current prokaryotic cells. In turn, as mentioned in Chap. 4, the emergence of regulation is the ground for many other fundamental biological capacities, such as adaptive agency and the capacity to establish many forms of symbiotic relations.

---

[28]This is because the particular sequence of its nucleotides is thermodynamically degenerated, in the sense that their order has no notable effect on the distribution of energy throughout the whole molecule. Instead, the alteration of nucleotide sequences in RNAs, and especially modifications in the sequence of amino acids in proteins, usually involve energy changes and therefore three-dimensional changes.

5.8 The Emergence of Darwinian Evolution

The appearance of a form of organisation endowed with two types of different components – some specialised in template functions[29] and others in catalytic functions – generated an evolutionary transition such that individual organisms bring about an unlimited variety of manifestations of their organisation, which is not subject to any pre-determined upper boundary of functional complexity (although they are subject to the energy-material restrictions imposed by a finite environment, by universal physicochemical laws and, moreover, by the principle of organisation; see Ruiz-Mirazo et al. 2008). The reason for this lies in their capacity, which is drastically enhanced by the action of inert templates, to indefinitely explore new innovations, coupled with their ability to incorporate and retain them.

The central feature of this kind of organisation is therefore twofold. On the one hand, any of the individual organisms can adopt a particular variant of the organisation. On the other hand, all of them, however different they may be, share a basic common organisational regime (as described above), whose long-term preservation depends precisely on this capacity for continuous, unlimited variation of the type of metabolism (i.e. closure endowed with regulation). This is because no matter how different the environmental conditions are, a new adapted variant of organisational closure may eventually be found.

We call the autonomous organisation described in the previous section "genetic templates-based organisation" (GTBO). Let us describe the different elements and relations underlying this new form of preservation through evolution (Moreno 2007).

First, GTBO requires a set of individual organisms (i.e. autonomous systems) capable of template-based reproduction and whose organisation includes two complementary polymers. GTBO, as explained earlier, is necessary to a full exploration of the sequential space in order to find new catalytic functions.

Second, these individuals generate a web of interactions. As Bedau (1996) has pointed out, a significant aspect of the environment to which any given organism must adapt is the set of all other organisms with which it interacts.

> So, when a given organism adapts and changes, the evolutionary context of all the other organisms changes. Thus, even without an externally changing environment, adaptation can be a co-evolutionary process that internally changes the selection pressures which shape adaptation, thus making open-ended adaptive evolution an intrinsic property of the system (Bedau 1996: 339).

In this sense, the evolutionary system includes an ecological dimension. As explained in Sects. 5.6 and 5.7., the templates-based reproducing individual

[29] In what follows, we will sometimes refer to this kind of specialised templates as « genetic ». This terminological choice is made to bring our usage into line with standard usage in the scientific literature; in turn, it does *not* imply an interpretation about their nature and role going beyond that which is explicitly provided in these pages.

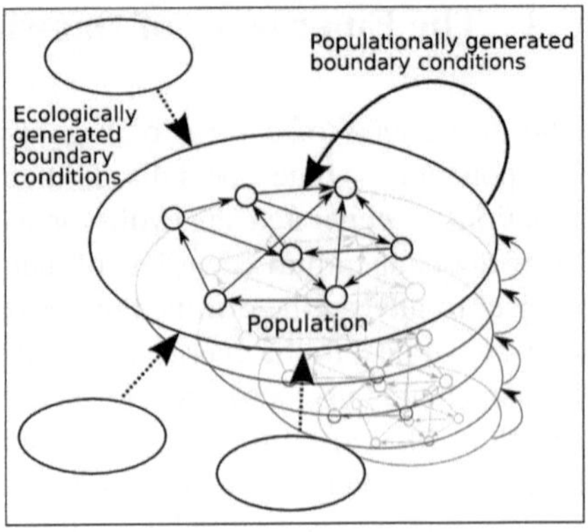

Fig. 5.3 The evolutionary system (Source: Moreno (2007))

organisations create and support a more comprehensive historical and collective domain, in which natural selection preserves the functional traits that provide some advantage in relation to the network of interactions at work between the organisms.

Third, this network is populational, and is so for two reasons. First, it requires a critical mass of individuals sharing the same specific form of organisation, as well as a certain degree of variability with buffering (within which selection is generated); second, it requires the existence of different kinds (species) of individual organisms. Through evolution, a diversity of species will be generated such that the boundary conditions for the self-maintenance of a given species start to depend on the interactions with the others (competitive and/or collaborative relations). In particular, the different reproductive rates will depend on the relations between species (and between individuals of the same species).

Accordingly, the evolution of the network is such that the species change according to the transformations of the boundary conditions affecting both the self-maintenance and self-reproduction of individual organisms. In turn, individual organisms are the result of interactions at the populational and ecological scales. The variation possibilities of both individual organisms and populations are, in principle, open.[30] This system of relations constitutes the core of the kind of evolution we are familiar with today, namely, Darwinian evolution (Fig. 5.3).

Through Darwinian evolution, an endless process of creation and preservation of organisational innovations takes place. Although there are some restrictions that apply to this process (organisational principles, body plans, internal laws of self-organisation, etc . . .), these very restrictions can also act as a set of constraints that enable the emergence of new structures and relations, thus allowing new forms of increasingly complex organisations.

[30]Until radically new forms of organisation (societies, technologies, etc.) emerge, thus transcending the fundamental biological organisation.

The genetic templates-based organisation that fundamentally grounds the simplest forms of present-day life – and, therefore, from our perspective, grounds the realisation of biological autonomy – was also qualitatively different from all its predecessors in terms of its long-term preservation. Whereas during prebiotic evolution, successive forms of organisation erased previous ones, once GTBO appeared, it would not only be preserved but would also become the condition of possibility for any further and more complex organisational step.

5.9 Is Darwinian Evolution Open-Ended?

The origins of autonomy, as we have seen all through this chapter, lie in a long-term historical process involving different steps. These steps are usually called "prebiotic evolution", that we have characterised here as pre-Darwinian evolution. During pre-Darwinian evolution, each new form of organisation erased the previous ones that had brought it forth, due to its superior robustness, efficiency, and capacity for long-term preservation. Darwinian Evolution, in turn, opened up a new era in natural history, by ensuring at the same time the long-term maintenance of life and an unlimited adaptive diversity of its core organisation (Fig. 5.4).

As detailed in the previous section, Darwinian Evolution relies on the following key-features:

1. Individual organisms must be able to reliably reproduce their genetic templates-based organisation (GTBO), which admits an unlimited variety of forms (unfolded in time);
2. The effective variety of these forms depends on boundary conditions that are also determined at the populational and ecological scales that, in turn, depend on the interactions between individual organisms;

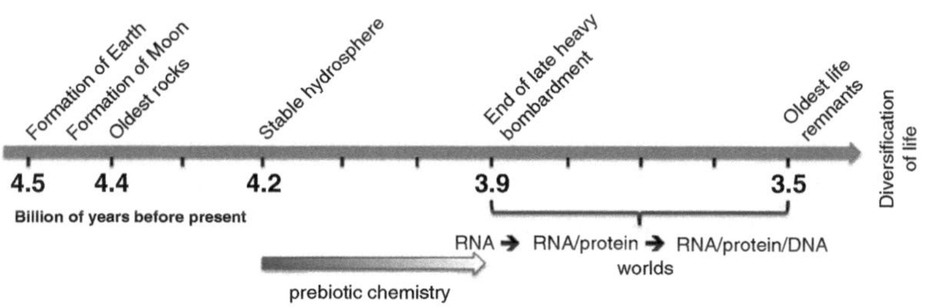

Fig. 5.4 Chronological scheme of the main steps in the prebiotic evolution of our planet (Credits: Juli Peretó)

3. The long-term preservation of the population of organisms is an open-ended process, not subject to any pre-given boundary of organised complexity[31] (Longo and Montevil 2014).

Let us now examine in some detail the very concept of "open-ended" evolution that we have been appealing to. What does it exactly mean? It seems to us that at least two interpretations are possible. On the one hand, open-ended evolution may refer to a process that brings about an unlimited *variety* of organisms sharing the same fundamental organisation (GTBO). Under this interpretation, we certainly think that Darwinian evolution is open-ended. On the other hand, the concept may indicate a process that would be able to attain unlimitedly higher *degrees* of organised complexity. The answer to the question whether evolution is open-ended in this sense is much less straightforward. As stated earlier, we are dealing thus far with a scenario in which the subject of Darwinian Evolution corresponds approximately to prokaryotic organisation, which is capable of displaying an unlimited variety of metabolic varieties. Among all organisms, indeed, prokaryotes are the most metabolically diverse and have some "exotic" ways to satisfy their needs. Yet, prokaryotic organisation is still subject to limitations as regards the increase of the degree of organised complexity (in particular, because of the lack of a compartmentalisation of the genetic material, see Mattick 2004). For example, no collective organisation of prokaryotes seems capable of generating an integrated and functionally diverse multicellular organism. In this respect, major biological transitions (as, for instance, the appearance of sex, multicellularity, the nervous system, language, etc.), and the subsequent realisation of higher degrees of complexity, require the fulfilment of further organisational conditions.

It is now quite unanimously accepted that the mechanisms of Darwinian Evolution (especially as far as phenotypic variation or plasticity are concerned, i.e., adaptability, generation of new functionalities, etc.) have themselves evolved (Conrad 1979; Wagner and Altenberg 1996; Kirschner and Gerhart 1998). One reason for this is the invention of new regulatory epigenetic mechanisms that interact with genetic, physiological and morphological systems, and may play a critical role in the transformation of the mechanism of evolution.

RNA editing, for instance, seems to play a key role in the evolution towards more complex organisms. As Gommans et al. (2009) have pointed out, genetic variation introduced through editing allows the exploration of sequence space that would be inaccessible through mutation, leading to increased phenotypic plasticity and providing an evolutionary advantage.[32] It is only over the last few years that we have

---

[31]Since any more complex form would not be preserved unless it were compatible with this organisational structure.

[32]With the invention of eukaryotic cell, (thanks, in particular, to the nucleation of DNA) organisms had the possibility for more elaborate regulation and processing of genetic information and, thereby, for a much more complex internal organisation than in prokaryotes. As J. Mattick (2004) has pointed out, in bacteria transcription and translation occur together: RNA is translated into protein almost as fast as it is transcribed from DNA. There is no time for intronic RNA to

become aware of the relevance and potential of these post-transcriptional regulatory mechanisms, mainly as a result of advances in comparative genomics and the realisation that non-coding DNA (or "junk DNA," as it was initially called) is pivotal for understanding plasticity in eukaryotes and later evolutionary transitions. There is evidence that the "microRNA repertoire" continued to grow during metazoan evolution, with very clear indications of this being found at the transitions to bilaterians, vertebrates, and placental mammals (see Hertel et al. 2006). In relative terms, the part of the genome that is responsible for these regulatory and epigenetic mechanisms keeps growing in importance, whereas the part responsible for core metabolic functions remains basically the same. The most recent surprise in this sense is the discovery that vertebrate genomes contain thousands of noncoding sequences that have persisted virtually unaltered for many millions of years (Mattick 2004). Furthermore, these sequences are much better conserved than those coding for proteins, a finding which was wholly unexpected.

So, the impressive morphological and physiological diversification of metazoan lineages (which implies a radical increase in both the variety and degree of the biological complexity) is based, once again, on the evolution of various regulatory processes that control the time, place, and conditions of use of the conserved core processes, which have modified their capacity to produce heritable phenotypic variation. However, all these innovations have not erased the basic mechanism of Darwinian Evolution; quite the opposite in fact: they require it. This is supported by the fact that more complex forms of biological organisation (e.g. eukaryotes, multicellular organisms, etc.) have not erased, but rather still critically depend on that minimal core, whereas the different infra-biological types of organisation, which appeared in the process of the origin of life, were soon "cleared away" by fully-fledged living beings.

In this fundamental sense, the conditions for bringing about the basic core of a biological organisation, i.e., the organisation of a population of prokaryotes and the multiscale (individual and historical-collective) system in which it results, are not only necessary but also sufficient for the long-term sustainability of life, because even if life had remained unicellular and "major" evolutionary transitions (Maynard Smith et al. 1985) had never arisen, the type of evolutionary pathway followed by living organisms would still be capable of producing unlimited functional diversity.

---

splice itself out of the protein coding RNA in which it sits, so an intron would, in most cases, disable the gene it inhabits, with harmful consequences for the host bacterium. In eukaryotes, transcription occurs in the nucleus and translation in the cytoplasm, a separation that opens a window of opportunity for the intron RNA to excise itself. Introns can thus be more easily tolerated in eukaryotes. In other words, the decoupling between transcription and translation permitted a much higher level of genetic regulatory control, which, in turn, would be required to increase the organisational complexity and plasticity of the whole cell (Taft et al. 2007).

5.10 Conclusion: Integrating the Organisational and Evolutionary Dimensions

In this chapter we have expanded the autonomous perspective by inserting its organisational framework in an evolutionary dimension. The general aim was to advocate both the possibility and necessity of bringing together (in accordance with other fields as Evo-Devo, see Laland et al. 2011) the theoretical legacy of the two different traditions in biology mentioned at the beginning of the chapter: the organisational tradition, focused on immediate or proximate causes; and the evolutionary tradition, leaning toward ultimate causes (Mayr 1961).

As a result, the autonomous perspective requires integrating two interrelated but different phenomenological domains. One is the world of physiological processes taking place in unicellular and multicellular organisms; the other is a domain of populational and inter-generational dynamics, occurring at a much wider spatio-temporal scale. The central issue addressed in this chapter is that these two domains inherently depend on each other and, in particular, autonomous systems cannot be generated independently from the historical-populational domain in which natural selection can operate. During evolution, individual organisms adopt very diverse forms of functional organisation; yet, they share a common organisational core, constituted by a form of organisational closure whose realisation and long-term preservation requires a set of almost inert specialised templates.

The autonomous perspective that we advocate, therefore, puts forward a picture of the phenomenon of life in which biological individuality cannot be severed from a wider collective organisation: as the individual organisation unfolds, it creates and supports a more encompassing historical and collective network, which in turn sustains and facilitates its evolution in a changeful environment. As Oyama (2002:164) has pointed out, evolution appears as "the derivational history of these organism-environment complexes". It is the interaction between processes taking place at different spatial and temporal scales that explains more adequately the powerful creativity of biological evolution.

From this integrated perspective, one of us has recently defined living organisms as autonomous systems with open-ended evolution capacities (Ruiz-Mirazo et al. 2004; Ruiz-Mirazo and Moreno 2009). On the one hand, autonomy covers the main properties exhibited by individual organisms, i.e. closure, metabolism, and agency. On the other hand, open-ended evolution captures the properties of life as an historical and collective phenomenon, i.e., as an entailment of reproductive cycles of autonomous individuals, bringing about the potential to innovate and increase biological complexity (always under the restrictions imposed, among other things, by organisational principles). The open-ended evolutionary capacity, hence, emphasises the fact that autonomous systems are inherently the result of an historical process, which relies – as we have described – on specific features and mechanisms going beyond the individual domain as the capacity for reproduction or reliable inheritance.

The fundamental connection between the organisational (individual) and evolutionary (historical –collective) dimensions of life explains why we have put so much emphasis on the structural requirements for the increase and preservation of biological complexity. Indeed, at the individual scale, one could hypothesise that the emergence of autonomous systems might possibly have occurred before and independently from the DNA-proteins world; yet, at the evolutionary scale, such hypothetical entities would presumably not constitute a genuine collective biological system insofar as they would not possess the resources for enabling its long-term sustainability. As stated at the beginning of this chapter, we do not need history to characterise what an individual organism is; we need history to understand where they come from, as a result of an evolutionary process, through which changes and variations can be preserved and accumulated, enabling a progressive increase of complexity.

6
Organisms and Levels of Autonomy

At the end of Chap. 4, we briefly mentioned that since the very beginning of life on Earth, organisms have established strong interactions (as opposed to weak ecological interactions) with each other, giving rise to several different types of stable associations. Unicellular organisms, which we took to be the prototypical example of autonomous systems, come together to form temporary bacterial aggregates, colonies, biofilms, and prokaryotic and eukaryotic multicellular ensembles. In turn, eukaryotic cells arise from symbiotic associations of prokaryotic cells and finally, colonial aggregates or more integrated societies establish groupings of multicellular systems with different degrees of cohesion. All these associations tend to occupy new niches and to increase the chances of survival of both the constituting units and the associations themselves as a whole. In certain cases, they even seem to behave as individual organisms.

One of the crucial issues discussed in the literature is to determine under what conditions these associations should be taken as fully-fledged organisms. The biological realm is full of examples of cellular ensembles or communities of cells, such as biofilms, slime moulds, lichens, sponges, mycelia fungi, clonal plants and colonial invertebrates, which may demonstrate some organism-like properties, but not all of them. In many cases, such composite multicellular systems dwell on the border between organismal and colonial behaviour, or between organismal and symbiotic relationships. It is therefore unclear in which cases they should be considered organisms, parts of organisms, or groups of organisms. As noted by Wilson (2000), assuming (as we do) that unicellular entities are organisms, the question would be: what sorts of multicellular systems meet equivalent requirements and can therefore be regarded as organisms? Actually, although we have pointed to unicellular entities as paradigmatic examples of organisms, it is more usual (or closer to our perspective as human beings) to think of highly evolved multicellular

---

This chapter relies on ideas previously formulated by Ruiz-Mirazo et al. (2000) and especially by Arnellos et al. (2014), from which several portions of the text are taken.

A. Moreno, M. Mossio, *Biological Autonomy*, History, Philosophy and Theory of the Life Sciences 12, DOI 10.1007/978-94-017-9837-2_6

systems as typical organisms (in particular metazoans, see Santelices 1999). Nevertheless, multicellular organisms represent a formidable challenge to any attempt to characterise or define them in precise terms, since cells have created many different kinds of collective entities over the course of evolution.

The contemporary literature shows that it is no easy task to determine which kind of organisation distinguishes "genuine" multicellular organisms from other forms of cohesive multicellular systems. Authors tend to offer a list of criteria (qualities and properties) that typify multicellular organisms, but often recognise that many exceptions exist. Sterelny and Griffiths' "spatial boundedness" (1999), Santelices' "unitary organism" (1999), Wilson's "paradigm organism" (1999), and the "functional integration" concept discussed by Wilson and Sober (1989) are examples of such criteria. Moreover, the criteria established in the literature are extremely heterogeneous; most are based on evolutionary considerations and even when they are conceived in organisational terms, they focus on very different aspects. So, although there is an intuitive grasp of the distinctive properties of organisms, there are always, as Clarke (2011, 2013) mentions, surprising cases of multicellular systems that force us to revise our criteria. Therefore, in order to make progress in this debate, what is required is a conceptual framework that, even if it does not completely succeed in clarifying the issue, at least provides us with the basic tools for interpreting most cases, including borderline ones, in a principled way.

In previous chapters, we have argued that individual biological organisms can be characterised as autonomous systems. When considering multicellular systems, the central question is whether the concept of autonomy developed so far also applies to such forms of multicellular organisation. What degree of integration and cohesion is required for multicellular systems to be taken as autonomous systems, and therefore as multicellular organisms? Supposing that we would agree that some multicellular systems indeed count as fully-fledged organisms, would it be in the same sense as for unicellular organisms? Furthermore, what is the status of the cells that constitute these different types of multicellular organisations? Are they still autonomous entities or just non-autonomous parts of an encompassing autonomous system[1]?

From the autonomous perspective, an organism is a regulated closed agential organisation that maintains itself while interacting with the environment. As we will see in Sect. 6.1.2, it seems reasonable to hypothesise that, in most cases, multicellular systems are self-maintaining closed organisations constituted by functionally differentiated parts (groups of cells) whose constituents (the individual

[1]The difficulty in applying the concept of autonomy to multicellular organisms was recognised by Maturana and Varela at the end of Chap. 4 of "The Tree of Knowledge" (1987), where they admit the problems involved in characterising multicellular organisms as "second-order autopoietic systems".

cells) are themselves closed systems. For example, a biofilm may contain many different types of microorganisms, e.g. bacteria, archaea, protozoa, fungi, and algae; each group performs specialised metabolic functions, and collectively they generate properties that emerge on free-floating bacteria of the same species. Accordingly, the biofilm constitutes a functionally integrated organisation that plays a causal role in the maintenance of the cells that actually constitute it. In our terms, biofilms realise a higher-level closure of constraints. Furthermore, multicellular systems are also integrated into ecological self-maintaining closed networks (see Chap. 4, Sect. 4.5 above), and often include deeply intertwined symbiotic associations.

In principle, as we argued in Chap. 1, closure constitutes a clear-cut criterion for marking the boundary between the system and its environment. In organisational terms, the set of constraints subject to closure constitutes the system, whereas all other constraints (and specifically those which have some causal interaction with the system) belong to the environment (as boundary conditions). When dealing with inherently intertwined multicellular biological systems, however, the question of the boundaries of closure may become much more complex, insofar as forms of strong (both intra- and inter-level) interactions between closed systems are considered. In spite of these difficulties, however, we do maintain that closure is a useful conceptual tool for identifying biological systems and, in particular, for distinguishing relevant levels of biological organisation. While we have previously discussed the realisation of different *orders* of closure (in relation to regulatory capacities), here we address the issue of *levels* of closure,[2] each level consisting of a set of closed constraints which is either made of constituents or included in an encompassing system, themselves realising closure.

At first approximation, the relations between levels of closure may consist in two different situations. In some cases, one can clearly distinguish between two or more distinct (and nested) levels of closure within the whole multicellular system. For example, in multicellular organisms, closure is realised by each individual cell on the one hand and by the organism on the other hand. Yet, it might be argued that there is no overlapping between the two levels of closure because individual cells do no exert functions that are subject to the higher-level closure. Only populations of cells (and, in ecosystems, populations of organisms) are subject to higher-level closure. In other cases, in turn, multicellular systems realise a kind of strong mutual dependence (symbiosis, for instance), which does not result in a sharp separation between the individual closures and the collective one. Within these systems, as argued by Ruiz-Mirazo and Moreno (2012), boundaries are not neat, though they can be established by "clusters" of mutual dependence. At some specific spatial

[2]As mentioned in Chap. 3, Sect. 3.2.2, each level of organisation can include one or more orders of closure, in particular if it possesses regulatory functions in addition to constitutive ones. Similarly, a given system can realise several levels of closure (and therefore of organisation), each of them including orders of closure. The conceptual distinction between orders and levels must be kept in mind to avoid confusion while reading the present chapter.

scale, in particular, many functions tend to be mutually dependent, such that one can identify discontinuities in order to establish the different levels of closure.[3]

Yet, as we already pointed out in Chap. 4, Sect. 4.5, when considering higher-levels of organisation, the realisation of closure does not involve as such the realisation of autonomy. As a consequence, the identification of higher-level closed organisations does not necessarily imply the identification of higher-level organisms. As a matter of fact, the encompassing multicellular organisation may perform a few functions (and simply maintain some relevant local environmental conditions for the different groups of autonomous agents that constitute it), while many others' functions are still subject to closure within the lower-level organisms. In the case of biofilms, for instance, different global properties such as density do play a role in the phenotypic shift of the bacteria. Moreover, because of the entanglement between the levels of organisation, it might be difficult to determine whether a specific function is performed by a given system or by an encompassing one, or by both of them. Thus, the main theoretical challenge consists in determining which of the hierarchically structured organisations that realise closure also meet the more demanding requirements for autonomy, and in precisely what sense.

In this chapter, we will not develop a comprehensive analysis of higher-level autonomy. Accordingly, we will not discuss all the implications of how multicellular organisms realise biological autonomy, and whether or not they are endowed with the very same organisational properties than unicellular organisms (notably with respect to distinctive regulatory and agential capacities). More modestly, our aim will be to make a first step in this direction by discussing some of the necessary conditions required for a multicellular system to be a *relevant candidate* as a higher-level autonomous system, and hence as an organism. In particular, we will focus on the kind of functional integration that a multicellular organism must exhibit. Our central claim will be that the *functional integration of multicellular organisms requires, as a necessary condition, developmental functions and, therefore, developmental constraints.*

The reason why we focus on development is that, to be such, a multicellular organism should not only be capable of reproducing each of its own parts but also its own collective organisation, which in turn requires some kind of developmental process, understood in a broad sense. In this respect, the analysis undergone in the following pages will rely on two ideas.

The first idea has to do with the impossibility of realising higher-level autonomy without crossing a sufficient threshold of diversity in the constitutive functional parts of the multicellular system and their reciprocal interactions. In short, sufficiently broad higher-level functional diversity is a necessary condition for functional integration that is strong enough. If the number of cell types in a multicellular system or the number of ways in which cell types contribute to the maintenance

[3] In this chapter, we do not offer a detailed account of the relations that might exist between entities located at different levels of closure. For more (conceptual and formal) details, see Montévil and Mossio (2015).

of the whole is too limited, there are not enough resources for a cohesive form of collective autonomous organisation to emerge. In our terms, minimal closure of constraints in the higher-level system is not enough: the number and diversity of organised constraints in the system has to be high enough to realise autonomy. The role of developmental processes is precisely to enable the generation of such a higher-level functional diversity that in turn requires developmental functions be themselves complex and various enough.[4]

A second but no less important idea concerns the centrality of *control* and, closely linked with this, of dynamic decoupling as a requisite for the type of organisation that may support a multicellular organism. For the question of generating functional diversity goes intrinsically with the problem of controlling it. Without higher-level control in particular, the simultaneous generation of rich functional diversity and high integration would not be possible. That is why intercellular control mechanisms stand out in all complex forms of development. In fact, they are so essential and pervasive in these systems that they effectively modulate the behaviour of the underlying metabolic units, i.e., of each of the cells that become part of the developing whole (their growth, differentiation, division processes), in the interests of the more encompassing *modus operandi*. Indeed, a very delicate and subtle balance between *intra*cellular and *inter*cellular dynamics has to be managed in the system, and this is simply inconceivable without the control exerted by higher-level functions.

In Sect. 6.1 we first briefly review the two main existing views on the concept of multicellular organism; we then argue that multicellular organisms require a set of developmental mechanisms governing cell differentiation as a necessary condition, enabling the establishment of a higher-level functionally integrated organisation. In Sect. 6.2 we examine in detail the developmental mechanisms of three specific multicellular systems and in Sect. 6.3, we discuss those three examples by analysing how their respective mechanisms subtend different degrees of higher-level organised complexity. Section 6.4 concludes the analysis, by focusing on the reasons why some of these multicellular systems might be legitimately said to realise higher-level autonomy, and therefore be qualified as multicellular organisms. Lastly, we briefly address the issue of the relations between levels of autonomy, specifically in the case of multicellular organisms composed by cells being themselves – by hypothesis – autonomous.

6.1 The Concept of Multicellular Organism: Evolutionary and Organisational Views

During the history of life, various forms of multicellularity have arisen independently in each of the kingdoms. Prokaryotes have recurrently demonstrated their capacity to establish multicellular systems with relatively simple architectural and

[4] Determining the precise threshold above which those critical transitions are triggered should be a fundamental empirical target of scientific research, and goes beyond the objectives of the chapter.

morphological features, made of just a few different cell types (Bonner 1999). Similar levels of complexity are observed in many cases of eukaryotic multicellularity (Bell and Mooers 1997). It is true, however, that the macroscopic and more integrated multicellular forms found in animals, plants, and fungi show a much greater functional complexity, as well as a remarkable variety of morphologies and underlying organisations. Hence, it is important to remark, first, that not all multicellular organisations show the same degree or kind of integration and cohesion (Kaiser 2001; Rokas 2008) and second, that multicellularity must be taken as a multifarious phenomenon that has emerged independently in the evolution of many lineages.[5]

Given the variety of forms and degrees of integration of multicellularity, there is a wide debate about the conditions at which a multicellular organisation should be considered a true organism (see for instance Santelices 1999; Perlman 2000; Ruiz-Mirazo et al. 2000; Pepper and Herron 2008; Queller and Strassmann 2009; Folse 3rd and Roughgarden 2010; Clarke 2011). The aim of this debate is to provide a definition of organism that could be used to deal with various open biological questions, insofar as organisms seem to be the implicit or explicit point of reference for basic biological concepts such as fitness, adaptation, generation, trait, phenotype, metabolism, lineage, development, natural selection, and evolution. In what follows, we will review some existing characterisations of multicellular organisms, which can be grouped into two main views. The first view conceives the concept of multicellular organism from an evolutionary perspective, as a unit of selection; the second deals with this concept from an organisational standpoint.

6.1.1 The Evolutionary View

As mentioned at the beginning of Chap. 5, evolutionary thinking conceptualises organisms as biological units to the extent that, by exhibiting variation, differential fitness, and heredity, they are entities on which natural selection acts.

In this view, in which the units of selection are what matters most, fitness and its maximisation are usually taken as the fundamental criteria for defining organisms (Gardner 2009). For instance, drawing on an analogy with a pocket watch, Gardner suggests that biological adaptation does not imply perfection or optimality, but rather *contrivance* (the property according to which "all of the parts of the organism or of the watch appear contrived as if for a purpose") and *relation* ("all of the parts of the organism or watch appear contrived as if for the *same* purpose" ibid, p. 861). He then argues that fitness maximisation is the key design principle that explains

[5]Multicellularity has evolved independently in prokaryotes and eukaryotes (Grosberg and Strathmann 2007). Although certain requirements for multicellular organisation (as cell adhesion, cell-cell communication, and cell death) already evolved in prokaryotes, complex multicellular organisms evolved only in six eukaryotic groups: animals, fungi, brown algae, red algae, green algae, and plants.

how natural selection solves the problem of adaptation, i.e. the "packaging" of parts into units of common purpose (be they organisms or watches) (Gardner and Grafen 2009). Therefore, according to Gardner, an organism is a whole whose parts are all under selection to maximise its own fitness.

In the same vein, Queller and Strassmann (2009) argue that the distinctive feature of organisms is adaptation, through which they demonstrate "goal-directedness" (p. 3144). They focus on the fact that an organism exhibits adaptations as a whole, and that these adaptations are not disrupted (at least, not significantly) by adaptations of the parts. In agreement with Gardner and Grafen, they suggest that:

> the essence of organismality lies in this shared purpose; the parts work together for the integrated whole, with high cooperation and low conflict (p. 3144).

High cooperation and low conflict between the parts of a system are therefore the relevant criteria for considering a system as an organism, and inferring that this whole is the locus of natural selection and adaptation (Strassmann and Queller 2010). These authors claim that "organismality" is something that needs to be explained in biology, as natural selection seems to condensate into organisms. Their approach complements the fitness maximisation view of Gardner and Grafen because they focus on actual rather than potential cooperation and conflict: "organisms should be defined as what they actually do" (p. 3144). They view germline sequestration as a capacity that evolved for controlling selfish mutations (i.e. decreasing conflict), and argue that more serious conflict happens when the requirement of "unicellular bottleneck" is violated, i.e. the fact that all cells of the organism come from one single, fertilised cell. Accordingly, they view plants as organisms as well, but see them as having somewhat higher conflict rates than animals due to their growth from multicellular meristems, which sometimes leads to actual conflict.

By defending a higher degree of cooperation than and a low degree of conflict between the interacting parts as the main criterion criteria for "organismality", Queller and Strassmann are not excluding the possibility that adaptations may take place above and below the level of the organism; rather, they argue that most adaptations will happen in discrete bundles, since the organism is, after all, the main focus of adaptation. In fact, these bundles of adaptations help identify organisms, because within each bundle almost all adaptations are directed towards a common end.

From a similar but more pluralistic perspective, Folse 3rd and Roughgarden (2010) emphasise that a definition based on the evolutionary concepts of fitness and adaptation would be preferable to one based on genetic and physiological characteristics. Following Maynard Smith and Szathmáry (1995), these authors claim that in an evolutionary approach to individuality, in which a new individual is considered as emerging from the interaction of previously independent ones, two main problems arise:

1. Selection operating at the lower level may be incongruent with selection operating at the higher level, and thus be fatal for the emergence of the new individuality;
2. Entities that were previously being reproduced independently, can now only reproduce *inter*dependently, as parts of a whole.

They then suggest what they call three "nested views of individuality", which should be jointly adopted to overcome these two difficulties. They call the first view "alignment of fitness", which stresses the importance of genetic relatedness and homogeneity, ensured by the unicellular bottleneck between generations in multicellular organisms. Basically, it is the idea that the organisation of cells avoids competition among themselves, so that the fitness appears as a collective property. The second view is called "export of fitness" and is based on the idea of germ-soma separation and the consequent division of labour between reproductive and non-reproductive tasks, which exports fitness from the lower to the higher level (see Buss 1987; Michod 1999, 2005).[6] The third view defines an individual organism as

> an integrated functional agent, whose components work together in coordinated action analogous to the pieces of a machine, thus demonstrating adaptation at the level of the whole organism (Folse 3rd and Roughgarden 2010: 449).

This third "functional concept" builds upon the "export of fitness", which transfers adaptations at the level of the whole organism, and therefore makes it the locus of fitness.

An important consequence of their tripartite and nested proposal is that "alignment of fitness" is not sufficient for individuality because, in this case, a multicellular organism would be equivalent, as Grosberg and Strathmann (2007) have suggested, to an ensemble whose parts (cells) stay connected after division.[7] Division of labour and functional organisation must be included to qualify a system as an individual from the "export of fitness"/"functional" point of view. As Folse 3rd and Roughgarden explain, the previous kind of multicellular ensemble (which just stays connected through generations without any cellular differentiation) would not demonstrate adaptation at the group level (the level of the whole), while the parts remain the locus of fitness. Therefore, the existence of the unicellular bottleneck is not sufficient for a transition to higher-level individuality: what is also required is an organisation of the constitutive cells that is complex enough to generate a functionally integrated multicellular unit. This is what we shall see next.

6.1.2 The Organisational View

The evolutionary view proposes a naturalised explanation for the design of organisms based on the mechanism of natural selection, analogous to the case of a watch. Kant had already used the same comparison in his *Critique of Judgment*, but in a rather different way. He noticed a fundamental *difference* between the

[6] More specifically, Michod (2005) has suggested that in a group of cells with complete germ-soma separation, the cell fitness of all cells will be zero, since none of the cells would be capable of both viability and reproduction (and the cell fitness is the product of them) although fitness at the group level could be considerably higher.

[7] As happens in all cases of multicellularity with an aquatic origin. See Bonner (1999) for details.

two: whereas the watch is formed by fixed components, fabricated beforehand and later assembled, the parts of an organism are formed for and from the others, some parts actually producing (and being in turn produced by) others. In our terms, organisms realise closure, while artefacts do not. Accordingly, while for the Darwinian tradition, the comparison between a watch and an organism – even regarding only contrivance and relation between parts – suggests an analogy, the organisational view requires an essential distinction.

As we explained in Chap. 1, an organism realises a closed organisation of constraints; its dynamic organisation plays a fundamental causal role in the generation of the constraints that actually make it possible. Closure, by definition, implies functional integration in the sense that the set of constitutive constraints exert mutually dependent functions that collectively maintain the whole organisation. Now, when dealing with associations of cells that not only become (temporary or relatively) cohesive systems, but may also turn into highly organised and functionally integrated entities, difficulties arise. Multicellular communities are made up of systems that are themselves functionally integrated, while at the same time they acquire some degree of functional integration and various degrees of interdependence at the collective level (Turroni et al. 2008). For instance, biofilms could be said to exhibit functional diversity in the sense that they bring together formerly differentiated groups of cells, performing several coordinated tasks (through the production of a common matrix, see Flemming and Wingender 2010; Ereshefsky and Pedroso 2013). In many biofilms, for example, there are groups of cells that belong to the multicellular entity only through the matrix provided by others. Just like biofilms, many other multicellular systems could also be considered, at least in a minimal sense, as organisationally closed systems.

Yet, the issue is that not all systems realising closure are eligible candidates for multicellular organisms. From our perspective, autonomy is the grounding of the concept of organism, be it unicellular or multicellular. Now, since we have argued that, in order to be considered autonomous, a system should realise a closed, regulated, agential organisation, the question is how and when these more demanding requirements are met in the multicellular domain. The organisational view should then clarify under what conditions multicellular closed organisations exhibit the relevant degree of functional integration for realising higher-level autonomy.

In this respect, the central remark is that, however different they might be, all highly integrated multicellular organisms are constituted by genetically homogeneous cells coming from one single fertilised cell ("germ cell"). In contrast to any artefact, or to weakly integrated multicellular systems, multicellular organisms result from a process of *differentiation* between their functional parts, and not from the *aggregation* of pre-existing entities. The main reason for this is that the forms of multicellularity constituted by genetically homogeneous cells, by enhancing integration, can considerably reduce intercellular conflicts. As Wolpert and Szathmary (2002: 745) have argued, only systems constituted by developmentally differentiated cell types are candidates as truly multicellular entities.

> It is advantageous for the unit of reproduction (the propagule) to be as small as possible (that is, a single cell), as the uniformity thus created will reduce the likelihood of conflict between cells. Mutation (...) will upset this uniformity, and selection against mutation may favour propagules of different sizes. Mutants that affect the organism but benefit the cell (such as those that lead to cancer) cannot be effectively selected out of large propagules, so their occurrence would favour a single-celled propagule. By contrast, uniformly deleterious mutants that affect the survival of both cell and organism can be successfully selected out of a multicellular organism, so their occurrence would favour propagules that are larger than a single cell (ibid.).

As these authors emphasise, only by meeting these requirements can multicellular systems evolve towards higher degrees of complexity:

> There are multicellular organisms, such as the cellular slime moulds, that develop by aggregation and not from an egg, but their patterns of cell behaviour have remained very simple for hundreds of millions of years. The evolution of more complex organisms increases the pressure to use an egg as a propagule (ibid.).

What is the link between differentiation and integration? Whenever multicellular organisms originate from germ cells, the generation of internal differentiation due to germ-soma separation entails some loss of freedom for single cells. More specifically, cells in a multicellular organism lose their totipotency through irreversible differentiation processes that make them apt to live only in a very specific environment, tightly surrounded by other cells, and therefore to contribute to the maintenance of the whole organism in a cooperative way. Therefore, the integration of functionally differentiated cells gradually emerges from early developmental stages onwards. For instance, inner cells depend on cells located at the physical boundary to obtain the material and energy resources required to carry out their own metabolism.

The connection between differentiation and integration has also been analysed by Buss (1987), who explains the origin of multicellular organisms from an evolutionary perspective as a unit of selection (from a similar perspective, it is also worth mentioning Michod 1999 and Bonner 2000). At the same time, he tries to integrate this evolutionary dimension into an organisational framework. Arguing that the germ-soma barrier is a derived evolutionary state, he shows how patterns in embryonic cleavage, gastrulation, mosaicism, induction, and competence arise as a consequence of the conflicting evolutionary interests of cells and the whole integrated multicellular entity. Buss explains that, in the evolution towards multicellular organisms, the germ line was initially not closed to genetic variations arising during the course of ontogeny. He studies the evolutionary emergence of homogeneous multicellular organisms as a competition between cell lineages to become germ cells, assuming that the unit of selection is the cell. In some organisms this evolution has produced homogeneity because germ cells are sequestered at very early stages of cell differentiation. Realising that there is a trade-off between the capacity for movement and the capacity for reproduction in single cells, Buss suggests that the appearance of gastrulation – where a hollow ball of cells is transformed into a multi-layered structure including diverse patterns of differentiated cells – was a crucial step in the origin of multicellular organisms. The idea was inspired by the observation that the cells of a metazoan can be either ciliated or prone to divide, but not both. In other words, the gastrula would be the "solution" to this problem, with the cells on the surface remaining ciliated while those inside lose their cilia, so they can divide. Through gastrulation, cells begin to live in a more specific and spatially-organised environment, where migrated cells are surrounded by still-ciliated ones, which stay at the periphery of the group and provide the material and energy inflow required for the proliferation of the internal cells.

Buss' account, being consistent with natural selection (since cells find a way to maintain themselves and proliferate), can then be said to show that differentiation and integration processes go together. Moreover, it shows that this must happen at a very early stage of development, in accordance with constraints that have been internally generated and should continue until a fully integrated multicellular system is formed. Buss' perspective is, no doubt, interesting. His strategy of accounting for the evolution of developmental architectures in terms of trade-off solutions for the conflict between selective pressures acting on cells and multicellular individuals points to what, in our view, are the fundamental questions for understanding the nature of highly integrated multicellular systems. However, his focus is mainly on the evolutionary origin of multicellular organisms rather than on the question of the organisational requirements for achieving multicellular organismality.

In contrast, our aim in the following sections is to examine, in some detail, the network of relations, mechanisms, and couplings that these associations of cells have to establish in order to achieve a higher degree of functional variety and integration at the collective level, to the point at which they can be considered multicellular organisms. In turn, the emergence of multicellular organisms requires what is usually called a process of "development".

As Wolpert and Szathmary have argued (2002: 745):

> The development of a complex organism requires the establishment of a pattern of cells with different states that can differentiate along different pathways. One mechanism for pattern formation is based on positional information: cells acquire a positional identity that is then converted into one of a variety of cellular behaviours, such as differentiating into specific cell types or undergoing a change in shape and so exerting the forces required for the formation of different structures. This and other patterning processes require signalling between and within cells, leading ultimately to gene activation or inactivation. Such a process can lead to reliable patterns of cell activities only if all the cells have the same set of genes and obey the same rules.

Furthermore, every state/phase of this developmental process should be sufficiently robust and reliable to be compatible with the requirements of natural selection (i.e., always above a minimal threshold of overall fitness). On this basis, we agree with Pepper and Herron (2008) that there is a type of "*positive feedback loop between the process of natural selection and the pattern of functional integration*" (ibid.: 626). Thus, the primary goal will be to provide a feasible explanation of the developmental requirements and characteristics of the mechanisms and organisation that give rise to such a positive feedback mechanism.

From this perspective, a necessary condition for the realisation of highly integrated closure – and possibly, higher-level autonomy – is that the system must include a specific class of functional constraints subject to higher-level closure, able to control the fate of the cells during the process of cellular differentiation. More specifically, this means that not only must the system possess constraints that are able to modulate *intracellular* epigenetic[8] mechanisms but also that they are also able to trigger off the generation of new developmental constraints during the process. Indeed, what matters for achieving multicellular organisms is the capacity to generate a high degree of phenotypic differentiation from genetically homogeneous cells. Under these conditions, not any form of higher-level control over development matters equally for the self-constitution and maintenance of the multicellular organism.

6.1.3 Multicellularity and Autonomy

Under what conditions may multicellular biological systems undergo the relevant complex collective process of ontogenetic development for getting higher-level autonomy? A sound answer to this question requires a characterisation of the endogenously generated cell-cell interactions resulting in the kind of functional integration of the systems under examination. One of the central challenges in this respect is to discern, as we will try to do below, what organisational level is ultimately "in charge" of the interactions (the individual cells or their collective organisation), paying attention to three specific, key features:

1. *Inter-cellular signalling mechanisms*, taken as one of the core aspects against which the size, diversity, and degree of sophistication of the interaction network can be assessed. This will be crucial for estimating the balance between *intra*- and *inter*-cellular constraints operating within the system as a whole.
2. *The plasticity, modularity, and robustness* of the network, trying to identify whether or not it includes higher-level functions. This in turn will provide an indication as to whether or not there is a set of interdependent constraints that functionally control the developmental process at the meta-cellular level.
3. *The degree of internal metabolic control over cell differentiation and cell division*. This will also provide an estimation of the extent to which the cell cycle is subordinate to the collective entity's global reproductive process.

---

[8]By the term "epigenetic" we mean processes and mechanisms by which a heritable phenotypic change is induced in the genetic system of a cell that does not involve a change in the nucleotide sequence of DNA (Berger et al. 2009). Epigenetic processes are basically the result of mechanisms allowing the selective activation of some genes and the inhibition of others. For example, DNA methylation or histone modification, which serve to regulate gene expression without altering the underlying DNA sequence. That is why epigenetic constraints affect the fate of the cells during development. Although there is no modification of the genome of the cell, epigenetic changes may remain through cell divisions for the remainder of the cell's life and may also last for multiple generations.

Taken together, these features provide a relevant measure of the degree and kind of control exerted by higher-level functions on the development and differentiation of individual cells. In turn, this gives an indication of the "taking over" of biological functions by the higher-level of organisation and, ultimately, of the degree of functional integration of the multicellular system as a whole. What matters from the autonomous perspective is that only those multicellular closed systems that have attained a sufficient threshold of collective functional integration are complex enough to realise higher-level autonomy. In particular, multicellular autonomous systems are those systems whose higher-level closed organisation includes the classes of functions required for autonomy, i.e. agential and regulatory. On the one hand, as described in Chap. 4, higher-level autonomy should include (in contrast with ecological organisations, see Sect. 4.5) agency, that is, the ability to deal with the environment *as an integrated (multicellular) unit.* On the other hand, regulatory higher-level functions are required, i.e. (Chap. 1, Sect. 1.8) *second-order* [9] constraints exerting their causal actions on changes of other constitutive constraints of the organisation.

As mentioned in the introduction, however, in this chapter we will not deal with the actual realisation of the interactive and regulatory capacities of multicellular organisms, nor even with the question of whether theoretical differences might exist concerning the way in which autonomy is realised at the different levels of organisation. Our focus here is on the control over development, as a general, necessary condition for attaining a sufficient degree of collective functional integration. In the next section, we will present and discuss three case studies in order to show how the appearance of increasingly integrated multicellular entities has required the appearance of increasingly complex strategies to manage the development of their internal functional variety and plasticity. In addition to illustrating several specific and empirically-grounded implications discussed in later sections, these cases help highlight the crucial importance of the kind of control exerted on cellular differentiation. In particular, we will examine the developmental processes of three multicellular systems: the cyanobacterium *Nostoc.punctiforme* as an example of a bacterial multicellular system; the green algae *Volvox.carteri* as an example of an early eukaryotic multicellular system; and the echinoderm *Strongylocentrotus.purpuratus* (a sea urchin) as an example of a metazoan multicellular system (Arnellos et al. 2014). As we will underscore, these systems exhibit substantial differences in the degree and complexity of the higher-level control exerted over development. These differences, in turn, underlie the differences in the functional complexity of the higher-level organisation. As a result, only one of these systems seems to be a candidate as a higher-level organism from the autonomous perspective.

It should be noted that, at both the prokaryotic and eukaryotic levels, there are many examples of multicellular systems (like biofilms) that are formed by

[9]The conceptual distinction between levels and orders of organisation is at work here. A given level of organisation, which is identified by the fact of realising closure, is a candidate as a level of autonomy if, among other things, it contains regulatory functions, subject to second-order closure.

aggregation, i.e. by the association of genetically inhomogeneous cells.[10] Yet, as we have argued, our underlying hypothesis is that only genetically homogeneous systems are relevant candidates as multicellular organisms.

6.2 Comparative Analysis

The three examples of multicellular systems examined in this section are relevant because they cover a wide range of configurations, while at the same time being highly integrated: they present a strong degree of contiguity (spatiotemporal neighbourhood) and, as we will explain in Sect. 6.3, they satisfy the criteria of alignment and export of fitness, exhibiting a high degree of collaboration. Accordingly, they might be taken as relevant candidates for multicellular organisms from an evolutionary perspective. Yet, as we will discuss, the analysis on how the developmental constraints operate in these three different multicellular systems may result in a different conclusion from the autonomous perspective.

6.2.1 Cyanobacterium Nostoc.Punctiforme

Nostoc.punctiforme is a multicellular genetically homogeneous system constituted by cyanobacteria, which form photosynthetic and diazotrophic filamentous organisations that obtain their energy from sunlight and their carbon from air and water, fixing molecular nitrogen as well. Initially, the cells are phenotypically homogeneous (all of them are vegetative cells), but then a developmental process takes place, leading to a phenotypic differentiation between two different types, photosynthetic vegetative cells and specialist nitrogen-fixing heterocysts.[11] Through this cellular differentiation, *Nostoc* can take energy from the sun and use it to make nitrogen compounds. The nitrogen products are then passed along to the photosynthesising cells.

The process of differentiation produces a semi-regular pattern of morphologically and metabolically different cell types. Several models have been proposed to explain this pattern of development (Kumar et al. 2010; Campbell et al. 2007). All models hypothesise that the pattern is ultimately determined by the action of a diffusible inhibitor produced by the differentiating heterocysts. The signal that

[10]An interesting case is *Physalia.physalis*, a highly integrated association of four specialised polyps and medusoids, whose constitutive parts can no longer disintegrate and continue living independently.

[11]Heterocysts are cells that specialise in nitrogen-fixing during nitrogen starvation. They fix nitrogen from dinitrogen (N_2) in the air in order to provide the cells in the filament with nitrogen for biosynthesis.

kick-starts development in all the vegetative cells is generated as a reaction to a nutrient limitation, namely, a lack of nitrogen. All cells in the filament detect the signal but only some of them respond to it, leading to a biased initiation process of differentiation. The cause of such biased initiation is not known, but it is assumed that it is associated with the physiological state of the cells (probably their position in the cell cycle at the moment the signal is detected). The nitrogen limitation triggers the activation of the global nitrogen regulator (*NtcA*) in all the cells that are in the appropriate cell stage. *NtcA,* in turn, at the same time, activates two different molecules, *HetR* (which induces the cell to differentiate into a heterocyst) and *PatS*. But whereas *HetR* operates intracellularly, *PatS* builds up as an intercellular gradient (as a molecular compound generated within the system), which diffuses rapidly among neighbouring cells. Diffusible *PatS* suppresses *HetR* and stops the differentiation of the neighbouring cell(s) as a heterocyst. If rapid diffusion drains *PatS* from the place of production (which is mostly the case), *HetR* synthesis is stabilised and the cell develops into a heterocyst. In neighbouring cells, the entry of *PatS* prevents the formation of *HetR*. In more distant areas, the diffusion of *PatS* may not be sufficient, so new centres of activation may be formed.

The crucial remark at this point is that no other cells have been found in the filament producing signals that act as intercellular constraints (i.e. inducing or suppressing cellular differentiation) on the cell that produced the diffusible *PatS*. Hence, it seems that the development of a differentiating cluster of cells into a single heterocyst (at any developmental site in the filament) operates, at a collective level, under the effect of a *single* constraint (*PatS* concentration). There seems to be no generation of other compounds/structures (i.e., no synthesis of any morphogen or some other kind of signal in the cells where *HetR* is suppressed by *PatS*) that act intercellularly on the phenotypic traits and organisation of the different cells that produced *PatS*, or indeed any other neighbouring cells.

Moreover, heterocysts undergo *terminal* differentiation, as they lose the ability to divide, because in this way they provide surrounding vegetative cells with combined nitrogen. In *Nostoc,* therefore, vegetative and reproductive functions are realised by the same type of cells. Furthermore, as explained by Christman et al. (2011), the transition from growth to nitrogen dependence (when heterocyst generation takes place) is not immediate. Given the limited number of developmental signals, cell division and cell differentiation cannot be modulated outside the core metabolic context. Consequently, the explicit dependence of the differentiation of a heterocyst on the vegetative cell life-cycle stage, and the terminal differentiation of heterocysts themselves, imply a mechanism of developmental modulation of differentiation that remains strongly coupled to the metabolic requirements of the vegetative cells.

6.2.2 *Volvox.Carteri*

Volvox.carteri is an eukaryotic multicellular system constituted by unicellular algae, which moves coherently towards the direction of light, and which has been frequently studied in attempts to understand the transition to eukaryotic multicellularity exhibiting cellular differentiation and complete germ-soma separation. This multicellular system has a developmental process that results in asexual spheroid adults with two cell types: large reproductive cells (*gonidia*) and small motile somatic cells. The coexistence of these two cell types, however, generates a problem, known as the "flagellation constraint" (Koufopanou 1994), namely the incompatibility between cell division and motility in photosynthetic flagellates. This incompatibility is problematic because all swimming photosynthetic organisms need to be motile even when they divide in order to maintain a position that allows them to efficiently use light for growth and division. *Volvox* solves this problem by differentiating a subset of cells in the anterior end into somatic cells that do not divide but continue beating their flagella, thereby providing the system with the capacity to swim. The rest of the cells (the germ cells) divide and produce progeny. Since germ cells directly become reproductive *gonidia*, *Volvox* exhibits a complete germ-soma separation.

How is this differentiation achieved? *Volvox* embryos first cleave and then divide asymmetrically to produce one large gonidial "cell initial" and one small somatic "cell initial" each (Kirk 1998). The first gonidial cells produce additional somatic initials at each division. The gonidial initials then temporarily stop any cleavage activity, while the somatic initials continue to divide symmetrically about three more times. At the end of embryogenesis, the volume of gonidial initials is about 30-fold larger than that of somatic initials. However, at this stage, cells differ only in size (Kirk 2005). Subsequently, the size of each sister cell leads to either a somatic or germline developmental process (Kirk et al. 1993). Thus, small cells develop as biflagellate somatic cells for motility, biosynthesis of the extracellular matrix and phototaxis, and large cells develop as non-motile, germ cells specialised for growth and reproduction. Asymmetric division plays a crucial role in *V.carteri* development, as has been extensively discussed (see e.g. Kirk 1998, 2005; Hallmann 2011 for details). Specifically, *Volvox* cells that are below a certain size threshold at the end of cleavage will differentiate as somatic cells, while cells above that threshold will differentiate as gonidial. In the case of a *gls* gene mutation, all cells will keep on dividing symmetrically, becoming somatic cells since they are too small to undergo gonidial specification. There is also another gene, *RegA*, which plays a crucial role for complete, stable germ/soma separation. It operates by repressing chloroplast biogenesis, thus preventing somatic cells from growing enough to trigger cell division. *RegA* mutants will follow the path of their unicellular ancestor, beginning as small flagellated cells and then re-differentiating as gonidia. By contrast, *lag* genes act in gonidia to prevent the development of somatic features, such as flagella and eyespots. Now, considering that all three genes (*RegA*, *gls* and *lag*) act *intracellularly*, and that the initiation of the somatic or gonidial developmental process is explicitly dependent on the size of each cell, it seems that cellular differentiation is achieved only by intracellular cell fate specification. In other words, the development of cellular differentiation in *Volvox* takes place independently of any intercellular signal produced by other cells (Nedelcu and Michod 2004) (Fig. 6.1).

As a result, in *Volvox* (even more explicitly than in *Nostoc*, since in *Volvox* there is a complete germ-soma separation) cell division remains either totally decoupled

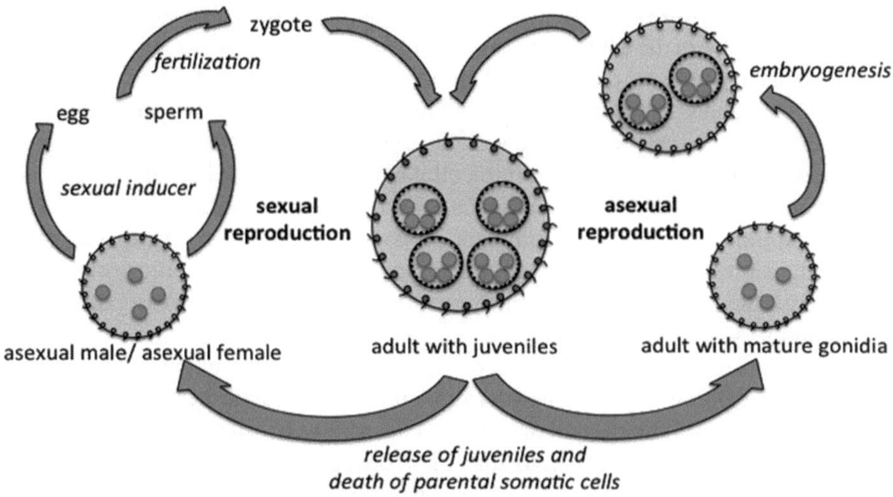

Fig. 6.1 Main steps in the development and reproduction of Volvox (*credits: Juli Peretó*)

(in somatic cells) or strongly coupled (in germ cells) to cell growth and global system reproduction. Consequently, *Volvox* does not present the flexibility required for further re-differentiation and growth in the somatic cells. The dissociation of cell division and cell growth does not occur through the asymmetric distribution of morphogens and germ-line factors during the asymmetric divisions in the cleavage phase, but rather by acting on the ancestral linkage of cell growth and cell division. While in *Nostoc* there is at least one intercellular signal, in *Volvox* developmental differentiation is entirely dependent on intracellular mechanisms and therefore it is strongly coupled to the core metabolic requirements of the processes of growth and division. Again, what is lacking is adequate higher-level control over development. As we shall see, this is precisely the difference with our next example.

6.2.3 *Strongylocentrotus.Purpuratus*

Strongylocentrotus.purpuratus, or the purple sea urchin, is a small invertebrate that belongs to the echinoderm phylum. Although it is a relatively simple metazoan, it has a very interesting developmental process leading to differentiated tissues and organs. Sea urchin embryos develop into free-swimming pluteus larvae.

At the beginning of development, pattern formation and cell differentiation in sea urchins employ two major mechanisms of cell fate specification (Peter and Davidson 2009, 2010, 2011):

1. The inheritance of maternal signals (structures playing the role of transcription factors) operating as intracellular determinants;
2. Intercellular signals (between cells of the same or of different lineages).

Initially the asymmetric distribution of the maternally provided signals along the major axes results in the establishment of domains of specific gene expression. This endows cells in different regions of the embryo with the capacity to send and receive intercellular signals. In this respect, the main difference with the two previous examples is that these signals lead to the variable expression of new sets of transcription factors, which by acting inter- and intracellularly, modulate the implementation and execution of *several different developmental processes*. According to these higher-level constraints, several developmental processes are initiated, stabilised or/and excluded, resulting in the spatiotemporal, timely production of specific proteins that characterise the state of different cell types, thus defining the overall organisational pattern in the developing embryo.

One of the most interesting aspects of sea urchin development is the mechanism of intercellular interactions that dynamically modulate key aspects of this development (Ben-Tabou de-Leon and Davidson 2007). These signals constrain the organisation of other cells, so that their developmental fate is appropriately specified and ensured. The results of the operation of intercellular signals are: (i) the initiation of the development of the endomesoderm; (ii) the timely separation between mesoderm and endoderm specification, and the initiation of mesoderm formation; and (iii) the timely separation between anterior and posterior endoderm specification, the initiation of their formations and the initiation of gastrulation. Let us briefly explain how all this happens.

Very early on, the intracellular operation of the maternally provided protein *β-catenin* creates a new signal, called *Wnt8*, whose intercellular operations result in a mutually constraining interaction between cells of the same lineage. In particular, *β-catenin* operates intracellularly, causing the creation of *Wnt8*, which in turn acts as a constraint on a neighbouring cell, in order that the nuclearisation of *β-catenin* in that second cell will be intensified, bringing about further production of *Wnt8*. This intercellular feedback mechanism ensures the continuous production of *Wnt8* across the lineage. This is essential for sea urchin development, since any disruption of that intercellular mechanism results in problematic specification of the skeletogenic and endomesodermal lineage (Oliveri et al. 2008). The intracellular operations of the increasingly nuclearised *β-catenin* will create another two intercellular signals: (i) an *early signal (ES)*, which is still undefined (Angerer and Angerer 2012), and (ii) a *Delta* signal, which will be used to drive mesoderm fate specification in the macromere lineage.

The indirect but mutually exclusive constraining actions between *Wnt8* and *Delta* operating intercellularly throughout the embryo's development are of particular interest here. The intercellular operation of *Wnt8* on the large micromeres induces *Delta*, whose intercellular operation results in: (i) the separation between mesoderm and endoderm developmental processes; and (ii) in the suppression of *Wnt8* in certain cells, permitting the creation of a new *Delta* signal in these cells. What happens in practice is that, as development proceeds, wherever *Wnt8* is generated in the endomesoderm, *Delta* is not, and vice versa (Peter and Davidson 2009). All these intercellular signals contribute to the precise activation of the mesodermal and endodermal developmental processes in space and time. Interestingly, this constraining process is much more indirect, as it is the result of other intercellular signals that operated several developmental stages back. This intercellular signalling continues throughout development, allowing the formation of tissues critical to the survival of the embryo.

In sum, the developmental process of the sea urchin is characterised by intercellular signals that constrain intracellular processes, which further specify or directly initiate the developmental fate of the respective cell lineages, and/or affect (by inducing or suppressing) the production of other intercellular signals. In turn, these signals will constrain the intracellular processes of other cells in the embryo. Accordingly, the type of development coordination occurring in sea urchins differs from that which takes place in *Nostoc* and *Volvox* in three main aspects:

1. In sharp contrast to the single intercellular signal for the development of the differentiating filament operating in *Nostoc*, and the purely intracellularly-determined specification of the two cell types in *Volvox*, the development of the sea urchin depends on *several* intercellular signals (as we saw, it depends at least on *Wnt8*, *Delta*, and others like *Wnt16* and *V2*).
2. In sea urchins, different types of relations (combinations) exist between the intercellular signals, resulting in different types of intercellular mechanisms.[12] In all cases the result is the creation of intercellular mechanisms that modulate the developmental process.
3. As a consequence, sea urchins seem to have the capacity for much more elaborate cellular differentiation, which is decoupled from the ancestral mechanisms of cell growth and cell reproduction. Cells preserve a degree of differentiation potential for several developmental stages, and the sequence of their biochemical changes and the timing of their division and/or migration are largely modulated by the combinatory application of past and present intercellular mechanisms operating on them.

Sea urchins modulate development and cellular differentiation through the operation of intercellular mechanisms that coordinate the fate of different cell lineages, while allowing new possibilities for cell differentiation; this gives rise to a new form of collective multicellular organisation. In the next section, we shall discuss the nature of the coordination of the three multicellular organisations and shall argue that the type of developmental modulation exhibited by sea urchins consists in a much richer functional variety, leading to more complex (possibly autonomous) higher-level organisations.

[12]One type is the intercellular feedback mechanism of *Wnt8*. Another type is the intercellular mechanism established by the indirect and mutually exclusive operations of *Wnt8* and *Delta*. A case of a highly combinatorial type of mechanism is the one at work for the separation between anterior and posterior endoderm formation, a process which is eventually established by the intercellular operations of other signals – *Wnt16* and *V2* – but which needs other inputs from the operations of other intercellular mechanisms during prior developmental stages.

6.3 Developmental Conditions for Highly Integrated Multicellular Organisations

The examples described in the previous section constitute three different kinds of multicellular systems, each of which could be pertinently described (at least from a phenomenological perspective) as resulting in a high degree of collaboration – and low conflict – among the parts. Accordingly, this would lead to fitness maximisation for the multicellular systems, and the new higher-level organisation could be identified with the capacity of a group of cells to demonstrate adaptation at the level of the whole multicellular system. From an evolutionary perspective, hence, all these systems might be taken as multicellular organisms.

From the organisational perspective, in turn, relevant fundamental differences exist between these systems. In spite of their common features, indeed, what matters (again, as a *necessary* condition) for the realisation of higher-level autonomy is the following twofold issue:

1. Whether, in these systems, organisational closure includes developmental constraints; namely, intercellularly produced functional signals modulating the fate of the cells during differentiation;
2. Whether these signals can trigger the generation, during development, of new similar control signals, in order to guarantee the establishment of substantial high-level functional variety and, in particular, of regulatory capacities.

With respect to both of these aspects, the developmental mechanisms of the sea urchin are significantly and qualitatively different from those of *Nostoc* and *Volvox*. The developmental process occurring in two latter cases is strongly coupled to the reproductive and self-maintaining intracellular lower-level processes. As a result, their number and complexity is severely limited.

As described in the previous section, *Nostoc* possesses only *one* signal constraining the intercellular dynamics in development. Moreover, the underlying mechanism of differentiation remains strongly coupled to the metabolic requirements of the vegetative cells. As a consequence, there is no development of further cellular differentiation, resulting in a functionally diversified and integrated high-level organisation. Things remain essentially the same with respect to the capacity for cellular differentiation in *Volvox*. Although in this case there is a much more elaborate process of development and reproduction resulting in a complete germ-soma separation, it seems that *no* signals act intercellularly to further modulate the dynamics of development. Last, but not least, the way this multicellular system maintains its germ lineage precludes any possibility of further re-differentiation and growth of its cells. Here again, the lack of control over the intercellular collective dynamics prevents any further development of cellular differentiation and higher-level functional complexity.

It is worth emphasising that *Nostoc* does realise second-order closure; it therefore possesses collective interactions that are necessary for its operational coordination (functional division of labour) and global behaviour. For instance, in *Nostoc* there is a rich exchange of metabolites between vegetative cells and heterocysts, which is necessary in order to meet the needs of the two cell types.[13] Yet – and this is the central point discussed in this chapter – this form of multicellular organisation does no generate a higher-level control subsystem operating on the developmental processes of the constitutive cells. In other words, in Nostoc, the multicellular system is unable to foster and support a process of development leading to the degree of functional differentiation required to get higher-level autonomy.

In the case of the sea urchin, things are substantially different. In sea urchins, the modulation of cellular development is based on several intercellular functional signals. These signals establish intercellular mechanisms that control the developmental process by triggering, activating, and suppressing intracellular processes responsible for the specification of the developmental fates of the respective cell lineages. As discussed earlier, different combinations exist of intercellular signals, which result in different types of intercellular mechanisms and, consequently, in qualitatively different kinds of developmental modulation. Subsequently, this allows for part of this set of intercellular signals to modulate intracellular processes that promote the production or suppression of other intercellular signals, which then constrain the intracellular processes of other cells, and so on and so forth (see Arnellos et al. 2014 for details).

These intercellular mechanisms constitute an endogenously created set of specific higher-level functions: through their constraining action, such a complex developmental process is effectively driven and the specification state of each cell lineage is spatiotemporally stabilised. As higher-level functions, these intercellular mechanisms are largely decoupled from the intracellular processes of the constituting units (because, among other things, their characteristic time scales are different; see Chap. 1), and can be varied without disrupting those more basic intracellular processes. At the same time, they act on the cellular epigenetic mechanisms, thus modulating their operations.

The specific organisation of the sea urchin can be usefully compared to different types of intercellular constraints that may also induce intracellular epigenetic changes, leading to some form of cellular differentiation. For example, some squids (*E.scolopes*) have a symbiotic relationship with certain bioluminescent bacteria (*Vibrio.fischeri*), which inhabit a special light organ in the squid's mantle. The bacteria are fed through a sugar and amino acid solution provided by the squid and, in return, "hide" the squid when viewed from below, by matching the amount of light hitting the top of the mantle. The light organ contains filters, which

[13]In the case of *Volvox*, the realisation of second-level closure is more debatable. Somatic cells achieve efficient swimming capacity that, thanks to their coordinated action, is beneficial to the whole system (and notably to reproductive cells). However, although there is coordination between reproduction (germ cells) and movement (flagellated cells), so that the network of cell-cell interactions results in a certain degree of functional differentiation, it remains unclear whether somatic cells could be said to depend on reproductive cells in the precise sense of "dependence" discussed in Chap. 1. Accordingly, the claim according to which *Volvox* is a multicellular organisation cannot be taken for granted, and would deserve further investigations.

may alter the wavelength of luminescence, making it closer to that of moonlight and starlight (McFall-Ngai 1999). In this symbiosis, the development of both the bacteria and the epithelial cells of the squid are modulated by each other. In particular, the bacterium *V.fisheri* induces several changes in the development of the squid's light organ, which lead to the loss of some superficial fields of the squid's epithelial cells; in turn, the squid induces two important developmental changes in *V.fisheri*, i.e. the loss of their flagella and the decrease in their cell volume. Yet considering that the bacterium has a prokaryotic genetic machinery[14] and that the type of cells responsible for the squid's intercellular developmental signalling network are completely different[15] from *V.fisheri*, the participation of *V.fisheri* in the squid's intercellular epigenetic network is severely limited, insofar as it does not have the capacity to significantly alter the network, in order to generate enough higher-level functional differentiation. From the organisational perspective, therefore, the symbiosis between the squids and *V.fischeri* does not set the conditions for higher-level autonomy. This suggests that the role played by symbiotic bacteria in the development of certain metazoans, although surely important (think of the human gut, for instance, as explored in Turroni et al. 2008), it does not succeed in reaching the degree of collective functional complexity of the multicellular systems constituted by genetically homogeneous eukaryotic cells.

In contrast, the genetically homogeneous (and epigenetically differentiated) eukaryotic cells in the sea urchin – and in the vast majority of metazoans – can participate in much more complex developmental functional interactions (e.g. leading to the formation of tissues and organs).[16] What the case of the sea urchin illustrates is the invention of a higher organisational level (an "intercellular epigenetic network") that enables the precise inter-level control of interactions

[14]For instance, *V.fisheri* has neither the ability to generate metabolically decoupled signals, such as *Delta* and *Wnt8*, nor the appropriate receptor mechanisms for their intercellular action.

[15]Eukaryotic epigenetic mechanisms are much more complex than prokaryotic ones because in the latter case, the processes of transcription and translation are operationally separated (see Chap. 5). Eukaryotic epigenetic control occurs even before transcription is initiated, and therefore in eukaryotic cells epigenetic mechanisms can control gene expression at many different levels. This means that intercellular signals modulating eukaryotic epigenetic cells can induce much more diversified effects.

[16]Although plants share many of the requirements so far described for developmental modulation, we centre our analysis in metazoans, because plants are multicellular systems based on cells with walls. Now, as Gerhart and Kirschner (1997) have pointed out, the loss of the cell wall in some unicellular eukaryote ancestor was also a very important factor in the appearance of rich cell differentiation in multicellular systems. "One development of great importance for future metazoan multicellularity was the loss of the cell wall in some unicellular eukaryote ancestor. The lack of a cell wall (...) permitted the ancestors of animal cells to interact directly with each other through apposed plasma membranes, to adhere to each other, to crawl on surfaces, to differentiate into complex shapes, to engulf other cells by phagocytosis, and to engage in junctional communication with other cells. Cell adhesion and junctional communication are characteristics of the formation of epithelia and the segregation of an internal milieu, which are found in all metazoans" (p. 11). See also footnote 18.

between functionally differentiating cells. In turn, this control over development enables, as claimed, the emergence of a much richer functional diversity, combined to a high degree of integration.

The relevance of higher-level developmental functional mechanisms in the case of metazoans is found in the complex problems inherent in the generation, maintenance, and reliable reproduction of their organisation. Their constituents are cells that already have a genetically instructed metabolism, expressed in different phenotypes. Therefore, the multicellular organisation must modulate cell growth, cell differentiation, and cell division, so that its constitutive identity (or at least some key aspects of it) is specified and coordinated by a self-generated developmental process.[17]

Metazoans are tightly integrated systems that constitute modifications, or redefinitions, of their meta-cellular organisation. These increasingly complex forms of organisation are based on deep-rooted changes at the developmental, body-plan level. In turn, developmental plasticity is possible, among other things, because of the regulatory possibilities offered by many animal-specific genes (e.g. homeotic genes) which, combined with different levels of RNA editing processes, expand enormously at these stages (Mattick 2004), and give rise to the generation of elaborate intercellular communication and adhesion devices, complete germ-soma separation and its integration in further cellular differentiation, sexual reproduction, etc.

In sum, the specific case of the sea urchin's development exemplifies a kind of higher-level control over development (widespread in most multicellular animals[18]), which sets the conditions for the realisation of a rich domain of functionally integrated higher-level organisations.

[17]While many other functional constraints may also contribute to the constitutive processes and maintenance of the whole multicellular entity (i.e., symbionts, indirect action of other organisms, etc.), they do not belong to the same level as those we have studied (namely, those constraints which regulate epigenetic mechanisms of cellular differentiation and which are decoupled from metabolic-interactive processes).

[18]Although multicellular plants have their own developmental processes too, it is undeniable that metazoans' development has achieved a higher degree of complexity. There seems to be a number of different reasons for this. First, unlike animals, plant cells do not terminally differentiate, remaining totipotent, often with the ability to give rise to a new individual plant. While plants do utilise many of the same epigenetic mechanisms as animals, such as chromatin remodelling, it has been hypothesised that plant cells do not have a "memory" and reset their gene expression patterns at each cell division, using positional information from the environment and surrounding cells to determine their fate (Costa and Shaw 2007). Second, the loss of the cell wall, already mentioned in footnote 16, seems to be an additional condition contributing to the enabling of the unfolding of the functional potentialities of cellular differentiation when building a complex integrated multicellular organism. See also Caroll (2001).

6.4 Towards Higher-Level Autonomy

In Chap. 4, we developed the notion of minimal autonomy, which provides the conceptual framework for characterising unicellular organisms as prokaryotic and eukaryotic cells (Ruiz-Mirazo and Moreno 2004).

The background question of this chapter is to what extent can this concept of autonomy be applied to a multicellular system. According to our definition, any entity that achieves regulated agential closure should be considered an autonomous system, and therefore an organism. Yet, the situation changes here, because we are dealing with systems whose functional parts are themselves constituted by autonomous entities (living cells). Indeed, in some forms of collective associations, the constitutive autonomous units may be more integrated and cohesive than the multicellular system itself. In other cases, instead, the global multicellular organisation becomes more complex, functionally diversified, and cohesive than that of its constitutive units. In many cases, therefore, it might be difficult to determine what level of organisation is ultimately responsible for the production (and control) of the constraints that drive the behaviour and interactions between the constitutive cells, in such a way that the whole set becomes a cohesive, self-maintaining agent.

In this chapter we have focused our study on the "internal" dimension of the problem, leaving aside the question of agency. As we have explained, development plays a crucial role in the realisation of higher-level autonomy. In particular, it is our contention that higher-level autonomy requires a high-level closure, including a set of developmental constraints, that is complex enough to ensure the generation of the adequate degree of high-level functional diversity. In contrast to what happens in the cases of systems as *Nostoc* and *Volvox*, the global multicellular system must produce a set of second-level constraints that functionally harness the developmental processes of each of the parts. If the collective entity meets these additional requirements, then it constitutes a relevant candidate as an autonomous organisation.

As argued in the previous section, not all multicellular systems constituted by genetically homogeneous cells exhibit the same type of functional organisation. Multicellular systems like *Nostoc* and *Volvox*, for instance, do not possess the capacity to exert sufficient higher-level control over the epigenetic dynamics taking place at the lower level. In some cases (*Volvox*) there is a total absence of intercellular signals constraining the developmental proceses. In other cases (*Nostoc*) they do implement minimal intercellular mechanisms that have a constraining effect on intracellular dynamics, but these are not diverse enough to open up a new functional and hierarchical domain that could lead to collective autonomous organisations. This is why *Nostoc* and *Volvox* do not meet the requirements as second-order autonomous systems.

By contrast, sea urchins possess an operationally closed combination of different types of intercellular mechanisms that control the epigenetic intracellular processes, so that cellular differentiation is enhanced and immediately channelled. As a result, sea urchins possess the relevant degree of higher-level of functional variety and integration: in particular, they seem to possess higher-level regulatory functions

(Peter and Davidson 2009, 2010, 2011). According to our definition, hence, they do might comply with one of the requirements for higher-level autonomy. And if, moreover, sea urchins were shown to possess an integrated form of agency,[19] they should be considered as multicellular autonomous systems.

There is one final issue to address before concluding. The claim according to which multicellular organisms realise *higher-level* autonomy supposes that those systems are made of components which are themselves (lower-level) autonomous. As mentioned in the introduction, this might be questioned, insofar as the constitutive units (namely, their unicellular parts) are very heavily constrained by the encompassing organisation. As we discussed in the case of the sea urchin, for instance, cells in the different cell lineages need to adapt their characteristics (the initial pluripotency of all blastomeres) to serve the multicellular organisation. Consequently, they undergo irreversible differentiation processes that make them apt to live only in a very specific environment, tightly surrounded by other cells. Therefore, not only do they become heavily dependent on each other, but also undergo deep-rooted organisational changes that allow them to form tissues and organs (something which requires a qualitatively different degree of collaboration, beyond metabolic or associative/aggregative exchanges). Accordingly, it might be argued that when unicellular systems become a part of multicellular autonomous systems, they are no longer autonomous.

In our view, however, this conclusion is not compelling. The fulfilment of the highly specialised functions of the multicellular organism seems to require a type of unicellular entity that, from the point of view of its internal organisation, still meets the requirements for autonomy (in particular, it continues operating through its own intracellular regulatory mechanisms, retaining a certain degree of epigenetic plasticity and even maintaining interactive functions), even though it can maintain itself only within the boundary conditions generated by the higher-level intercellular mechanisms. In multicellular autonomy, each cell maintains its own identity, based on a closed network of chemical reactions that generates its constitutive and agential dimensions. The control exerted by the multicellular organisation on the individual cells is restricted by the need to preserve the metabolic coherence and minimal threshold of epigenetic plasticity of the unicellular units. In a word, it seems that multicellular autonomy does require unicellular autonomy.[20]

---

[19]Elsewhere (Arnellos and Moreno in press) one of us has argued that only eumetazoans *do* meet the criteria for being considered as multicellular agents; actually, such agential capacities are deeply related to the kind of development described in the case of sea urchins in the preceding pages.

[20]The actual characterisation of such nested levels of autonomy might not be an easy task. To mention again the issue of agency, it might be quite difficult, in some cases, to locate specific agential capacities at the relevant level of organisation. Deciding "who is the agent", so to speak, may therefore require fine-grained analyses when dealing with multicellular systems whose components are themselves autonomous.

7
Cognition

In the history of life on Earth, some organisms have developed highly complex interactive capacities while others, which have evolved within specific viable niches, have not. As a consequence, living systems exhibit a wide variety of interactive capacities, ranging from fairly simple to extremely complex. For instance, animals and plants exhibit very different types of agency and, roughly speaking, we tend to associate more complex agency with the former than with the latter type of living system.

The general issue that we address in this chapter is whether and how, from the autonomous perspective, *cognitive* phenomena can be understood as a specific and highly complex class of interactive capacities, stemming from the evolutionary complexification of agency.

There is wide disagreement in the contemporary literature regarding what kind of interactive capacities should be qualified as "cognition". This disagreement seems difficult to reconcile because the different views are formulated as (or grounded on) intuitive definitions. Traditionally, the authors that might be included in the autonomous perspective have tended to defend the view that because adaptive autonomous agents are capable of "enacting a meaningful world", all autonomous agents (and therefore, all living beings) would be *ipso facto* cognitive agents, at least in a minimal sense (Maturana and Varela 1980; Maturana and Varela 1987; Bourgine and Stewart 2004; Stewart 1996; Thompson 2007). According to these authors, then, the adaptive behaviour of minimal organisms (such as bacteria) is already a cognitive phenomenon.[1]

Some of the ideas exposed in this chapter are taken from Moreno and Lasa (2003), Moreno and Etxeberria (2005), Barandiaran and Moreno (2006) and Barandiaran (2008)

[1]It is worth noting that there are other authors belonging to the autonomous perspective, such as Hooker, Bickhard, Christensen, and Collier, who do not identify life with cognition. In very broad terms this group seeks, as we do, to characterise cognition in more restrictive organisational terms than just the possession of agency. However, whereas they look to increased behavioural capacities

A. Moreno, M. Mossio, *Biological Autonomy*, History, Philosophy and Theory of the Life Sciences 12, DOI 10.1007/978-94-017-9837-2_7

In contrast, more traditional research programmes in Cognitive Science and, in particular, Artificial Intelligence, have mainly conceptualised cognitive functions as high-level capacities resulting from symbol manipulating programmes in abstract contexts (Newell 1980). In this case, it is only possible to talk about cognition when we find fully-fledged forms of rationality or, at least, human-like linguistic capacities. More generally, for many cognitive scientists, human (or hominid) cognitive capacities serve as a model for identifying what is or is not cognitive. This second position has the advantage of dealing with a distinctly cognitive phenomenology by focusing on high-level interactive skills. In this view, therefore, only humans (and occasionally certain mammals to the extent that they show a strong resemblance to human cognitive capacities) can be considered "cognitive agents".

Yet, both these views face serious problems when adopted for developing research programmes aimed at improving our understanding of the phenomenon of cognition within a biological framework. In the second case, this is because by leaving aside the whole evolutionary trajectory that gave rise to these capacities, it is more difficult to understand their functional origins or their relationship with the whole organism. In the first case, the weakness is that by dissolving cognition in broader biological phenomena, it is difficult to understand the nature, role, and more particularly, the evolutionary history of cognition as a specific phenomenon (Moreno et al. 1997)

We believe that substantial progress can be made by changing the way in which the problem is formulated. In this chapter, then, we will first address the issue of the relationship between agency and cognition from an evolutionary standpoint by focusing on the processes that, throughout the history of life, have led to the emergence of cognitive capacities. We will begin with more simple agential ones and, instead of directly trying to answer the question "what is cognition?", we will try to provide some elements of response to the question "how has cognition evolved from simpler forms of agency?". Thus, our strategy reframes the problem: instead of focusing on straightforwardly defining the *boundaries* of cognition, we will first discuss the *evolutionary transitions* towards more complex forms of agency.[2] Only at the end of this discussion will we adopt a position in the ongoing debate, putting forward a categorical definition of what makes some interactive capacities "cognitive".

---

as the primary discriminating dimension, our focus is on understanding the increasing functional capacity and complexity of the underlying biological organisation. The approaches are compatible with one another in principle. Yet, because of the specificity of the constraints imposed by the embodiment of functional capacities, we consider our own approach the more fundamental and do not pursue behavioural alternatives here.

[2]This is not to say that an increase in complexity implies an evolutionary advantage; viruses and bacteria are "as adaptive" as, for instance, large primates. We just assume that evolution has explored new forms of organisation, and some of these are more complex. Increase in organismic complexity during the course of evolution can be explained by the fact that, starting from a simple base, there was nowhere else to go (Gould 1988, 2002).

We expect this way of addressing the problem to help characterise cognitive abilities by contrasting them with the broad biological mechanisms of adaptive agency. If we were able to determine the basic milestones of the chain of causal events that led from the most basic forms of adaptive agency to the first traces of mind and consciousness (something that, as we will discuss, seems to have occurred at some stage of vertebrate evolution), we would be in a better position to replace intuitive criteria with a more objective account of cognitive phenomena. Yet, such characterisation will at best apply to what can be labelled "minimal cognition", i.e. cognition as it appeared at the early stages of the evolutionary transition from agency, rather than its fully-fledged expressions in some classes of biological organisms.

7.1 Agency and Motility

Living beings show very different types of agency and generally speaking, we tend to associate cognitive phenomena with certain complex forms of adaptive interactions of animals (i.e. with complex behavioural agency, which involves motility) rather than with those of plants. This is for two reasons: first, because the former appear to be more complex; and second, because they tend to more closely resemble our own cognitive processes.

If one accepts this difference by hypothesis, the central questions are then: how do we explain the *specificity* of the evolution of animal agency, especially that of vertebrates? What has driven this process of complexification of agency in animals? And what connects, from an evolutionary perspective, motile agency and cognitive capacities? Recently, Christensen (2007) has addressed this question in the following terms: what determines significant variations in the increase in complexity of agency during evolution? As he points out, a theory of cognition should explain the evolution of sensorimotor organisation and behaviour in metazoa and be able to say "what it is that is under selection when cognition evolves". Christensen believes that the answer to this question lies in the set of factors that account for the evolution of what he calls "higher-order control" in the organisation of agency. He argues that this claim is consistent with the more general structure of the evolution of sensorimotor systems in vertebrates and, moreover, that this fact refutes the challenge presented by the advocates of theories of distributed cognition. In this chapter, we will elaborate on Christensen's perspective.

In the following subsections, we will focus on the connection between the evolution towards more complex and efficient motile agency and the organisational requirements that multicellular organisms must meet. In particular, for reasons that we will discuss later, the appearance of the nervous system stands as a crucial step towards the further emergence of cognitive capacities.

7.1.1 The Relationship Between Multicellular Integration and Behavioural Agency[3]

During evolution, the formation of multicellular systems offered several selective advantages.[4] Besides size itself, which in certain cases gives access to a new niche, multicellular systems can create new interactive functions, thus increasing the fitness of their (uni)cellular constitutive entities. For example, the fact that some actions are performed by the multicellular system, instead of the unicellular one, may provide an adaptive advantage by breaking down certain food sources thanks to the collective excretion of enough hydrolytic enzymes, by increasing resistance to chemical substances (e.g. penicillin-resistant biofilms), etc. . . .

One of the most important and characteristic advantages of these multicellular communities lies in their capacity to deploy global, coordinated forms of behaviour (Shapiro 1998; Kaiser 2001). In particular, thanks to their capacity for *integrated motility*, certain multicellular systems develop new methods of obtaining food, better means of avoiding predators, higher capacities for predation, more effective means of dispersal, and access to resources and niches that are beyond the capacities of the respective isolated unicellular entities. Hence, these multicellular communities might be more able to adapt to their environments and increase the possibilities of survival of their constituting units, compared to what these units can achieve in isolation.

However, as was discussed at length in Chap. 6, not all forms of multicellular aggregation have the same degree of integration. This is important since the appearance of new and more complex forms of agency depends on the degree of integrated individuality that multicellular systems possess. As Hooker (2009) has pointed out:

> the emergence of multicellular organisms represents a massive expansion of both interactive capacity and self-regulation of that capacity and in this lies their rich adaptabilities that make them so successful. The focus of understanding such evolutionary functional change should thus be, not on finding repeated levels of the same stringency of closed cellular autopoietic organisation, but instead on the effective mastery of increased interactive

[3]Many of the ideas of this section are taken from Arnellos and Moreno (in press).

[4]The path followed by organisms to grow in size is multicellularity because cells seem to be unable to overcome certain size limits. According to Bonner, this has to do with energy considerations: "if one thinks of the rates of different chemical processes occurring within the cell, the distances needed for diffusion, the surface boundaries needed for isolating different chemical components of the motor, and so forth, all of these lead to the conclusion that there is an optimal size with sharp upper and lower limits, which is the size found in nature" (Bonner 1988: 61). For others (Vogel 1988), it has to do with the appropriate size for transmitting genetic products via diffusion. In fact, both arguments point to a similar problem, namely, the progressive difficulty of maintaining a molecularly based metabolic and reproductive organisation as size increases. (Moreno and Exteberria 2005)

openness. That mastery is achieved through increased self-regulatory capacity to modify, in situation-dependent ways, both the internal metabolic and external environmental cycles. (521–522)

As a matter of fact, even some classes of multicellular systems that exhibit a certain degree of developmental differentiation – especially plants and fungi – do not exhibit interactive capacities involving any great innovations in comparison to those of their constitutive parts. In turn, within the same domain of eukaryotic multicellular organisations, metazoa do have developed highly integrated and complex forms of motile agency. Let us consider and compare some examples.

M. xanthus, a prokaryotic multicellular system behaving as a single entity,[5] achieves a form of agency that increases the survival of its constitutive cells. The cells exhibit coordinated movements through a series of signals creating dynamic patterns in response to environmental cues, which facilitates predatory feeding. The cells move by gliding (a movement in the direction of the long axis of the cell) on a solid–liquid surface without the aid of flagella. *M. xanthus* has two genetically distinct systems for gliding, one of them is called *social motility* and involves the movement of cells in groups. Social motility is based on certain type of pili, extruded from one cell pole, and which adhere to a surface or to another cell. *M. xanthus* cells periodically reverse their direction of gliding and, within swarming groups, present different sets of movement depending on the absence or presence of a prey. In the former case, movement is quite random, but in the presence of a prey, there is a movement towards prey concentrations (Berleman and Kirby 2009). This provides the opportunity for detecting significant variations in the quantity of food resources without a significant change in the position of the cells. So, despite the very slow movement achieved by *M. xanthus*, this behaviour increases the chances of its constitutive cells remaining in close proximity to the prey.

Another interesting case is the eukaryotic multicellular system *volvox carteri,* discussed in Chap. 6. *Volvox* has a developmental process by which certain cells become flagellated and located on the outside of the system (which adopts a spheroid form), while others become sensitive to light. Consequently, the multicellular system as a whole adopts a light-searching behaviour. Moreover, the coordination of flagella beating between somatic cells provides *volvox* with the capacity for integrated movement (it can swim up to 2–3 m per hour[6]) and subsequently with the ability to actively look for environmental conditions with better luminosity. All these properties reinforce the whole system's capacity for survival. *Volvox* moves consistently towards the direction of light, because the flagella of its cells beat in synchrony. The coordinated mechanism enabling global motility works as follows: *Volvox* cells are arranged along the periphery of a spherical aggregation of cells, with their flagella pointing to the outer side; the flagella orient in specific directions and

[5]Actually, the constitutive cells of *M. xanthus* could live independently. When they do not find sufficient nutrients, they aggregate to form raised pigmented mounds, termed fruiting bodies.

[6]This is three times slower than unicellular paramecia, which move by means of numerous cilia beating in a coordinated way at rates of up to 2 mm/s.

beat synchronically to provide coordination in *volvox* movement; a large eyespot in each cell detects light and enables *volvox* to move in this direction.

The point is that the motile agency of both *M xanthus* and *volvox* faces several limitations. In both cases, the variety and flexibility of the behaviour of the whole entity is very poor, because the sensorimotor coordination of different cells is regulated biochemically and is heavily dependent on the underlying metabolic processes, thus achieving very low degrees of plasticity and modulatory capacity. Moreover, the small number of cell types does not allow the construction of an organisation that will ensure fast and versatile sensorimotor coordination, which precludes the diversification of its repertoire of movements (*volvox,* for example, can only swim towards the direction of light and at a very slow speed).

The situation is similar in the case of plants able to perform faster movements. At the end of Chap. 4, we have already mentioned the case of certain carnivorous plants, like *Dionaea muscipula* and *Aldravenda vesiculos*. These plants rapidly close their leaves when the sensitive hairs on the leaf lobes are triggered, and in this way they can capture small invertebrates. When a pray hits on the surface of the leaf, it very easily touches a hair. In this way, it triggers a kind of sensorimotor mechanism. When the hairs are stimulated, an action potential is generated that propagates to certain specialized cells, which in turn respond by pumping out ions that cause water to follow by osmosis. The mechanism is indeed very fast; it takes usually less than a second. Yet, this form of multicellular agency, though much faster than that of M *xanthus* and *volvox*, lacks flexibility and plasticity, and does not involve the whole body of the organism. Despite the fact that these multicellular systems possess much richer cellular differentiation than *M. xanthus* and *volvox*, the achievement of plastic and versatile integrated motility seems to require the fulfilment of additional organisational requirements.

By contrast, metazoan multicellular organisms can achieve qualitatively different forms of behavioural motile agency both in terms of velocity and plasticity. A jellyfish (which is a metazoa belonging to the phylum cnidarian), for example, moves by squeezing its body so that jets of water from the bottom are pushed out, which causes the jellyfish to be propelled forward. This movement is much more plastic (in addition to the diversity of their body movement when swimming, jellyfish also move their tentacles to gather food and sting potential attackers) and faster (one kind of jellyfish, the so-called "sea wasp", can reach speeds of 1.8 m per second; this is thousands of times faster than *Volvox*). Moreover, jellyfish already possess sense organs in the form of eyes and statocysts (Jacobs et al. 2007), which allow them to engage in targeted, precise behaviour. Jellyfish eyes range from simple eyespots and eyecups to relatively complex eyes with a lens. Extraocular photosensitivity is widespread throughout cnidarians, with neurons, epithelial cells and muscle cells mediating light detection. The aforementioned "sea wasp" (or "box jellyfish") has camera-type eyes with a cornea, lens, and retina (Kozmik et al. 2008), thus allowing the animal to avoid obstacles while swimming at high speeds.

At first sight, if one compares the three former cases with this latter one, it might be concluded that their behaviour is significantly "less efficient". Yet, if we consider the function of ensuring their viability, their behaviour is no less efficient than that of the jellyfish. The degree of complexity of a given behaviour is not per se related to its efficiency. In fact, what truly matters when focusing on the interactive functions that these multicellular systems perform (and especially in those cases in which agency is motility-based), is the relationship between these functions and the encompassing organisation. This relationship is two-fold. As we will discuss in the next section, only the appearance of much more complex multicellular organisations in metazoan evolution has enabled new forms of individuality and agency: the path towards complex forms of agency is inherently linked to that towards complex multicellular systems. In turn, the more the organisation of multicellular organisms is integrated, the more their behavioural agency should be faster, much more plastic, and efficient; otherwise, the latter could not satisfy the normative conditions of ensuring the maintenance of the constitutive identity of the multicellular entity as a whole. For example, a jellyfish could not obtain food and avoid predators without its very simple nervous system. In addition to its fundamental role in supporting the jellyfish's agency, the nerve net also plays an important role in the regulatory control of the animal's development, homeostasis, and reproduction (Arnellos and Moreno in press). Jellyfish nervous system possesses almost the full range of neurotransmitters, neurohormones, and non-neuronal hormones present in chordates or arthropods. Neuropeptides have been shown to systemically act like true hormones during development, homeostasis, and reproduction. In fact, reproductive and growth phenomena are under the control of neurohormones released by the neurosecretory cells (Hartenstein 2006). This is a very significant fact because the nervous system, as we shall see next, supports the regulation of both sensorimotor behaviour and internal constitutive processes.

Let us now give a closer look at the relations between organisation, integration, and behavioural agency.

7.1.2 *Organisational Requirements for Complex Behavioural Agency*

The appearance of higher-level organisms was a turning point for the enhancement of behaviour. At the multicellular scale, it becomes impossible to perform quick and versatile sensorimotor actions based on metabolism alone. There are two reasons for this: the enlarged internal distance between parts of the body, which need to be connected in very short laps of time (so that the organism can move fast); and the need to selectively modulate the organisation of connections in order to enable adequate sensorimotor correlations.

Therefore, if metabolic network plasticity were the only mechanism for accomplishing adaptive interaction and self-maintenance, the behavioural repertoire of multicellular organisms would be very limited. At the same time, multicellularity has permitted the emergence of fast interconnections among body parts quite independently from metabolically mediated processes, which avoids these limitations and opens up access to a new range of ecological niches.

How does multicellular organisation enable the appearance of efficient, plastic and integrated motility? The requirements are complex and only became possible with the advent of developmental processes that enabled the creation of much more integrated bodies, endowed with specialised structures (Arendt 2008; Keijzer et al. 2013; Arnellos and Moreno in press). In particular fast, plastic movement in multicellular organisms is only possible through muscles, which directly convert metabolic energy into mechanical energy independently of the continuous process of metabolic self-maintenance and morphological transformation[7] that the organism undergoes by means of cell growth and reproduction. Muscle cells evolved by assembling new variants of motor proteins for fast and slow contraction and by forming adhesive substrates that can withstand and counteract the generated contraction forces (Seipel and Schmid 2005).

In order to generate flexible and efficient motility, however, such cellular contractions had to be coordinated in such a way that they produce a globally integrated movement of the body. In turn, this whole musculoskeletal system requires the development of sense organs, because the complexification of behavioural agency also requires the capacity to detect increasingly complex environmental features. In fact, there is an association between the development of appendages and that of sensory organs (Jacobs et al. 2007). Last but not least, muscles and sense organs must be connected by a sufficiently fast and plastic network, namely, the *neural network*. Let us focus on this specific point.

Body movements, if supported only by a conductive epithelium, lack flexibility and efficiency. Given their anatomical structure as uniform sheets, epithelial tissues have obvious limitations for supporting precise and flexible muscle activity because accurate, long distance, targeted connections are not possible (Keijzer et al. 2013). This is why the appearance of neurons[8] (which, in turn, was possible thanks to the rich variety of cellular differentiation in eukaryotic cells) was so important: only these specialised, elongated cells with chemical transmission enabled precise and targeted connectivity between sensor and effector surfaces, over and above the more basic and diffuse conductive capabilities of epithelial tissue.[9] The development of the nervous system enabled multicellular organisms equipped with a musculoskeletal system to behave as a single, integrated entity. The requirement of a nervous system to support flexible and efficient motility at the multicellular scale is evident in the fact that, with the exception of sponges, which are almost sessile animals, all metazoa do possess a nervous system.

---

[7]Actually, this is a convenient simplification, because in fact the metabolic system sustains the neuro-muscular system; moreover, it also affects it (e.g. when exercise, skeletal growth, or injury demands an adaptation of the neural system), and is affected by it, directly via the neuro-endocrine system and indirectly via directing behaviour (e.g. forcing exertion until organs start to dissolve).

[8]Neurons are different from other cells in that they are capable of forming branches that are interconnected through plastic electrochemical pathways and are capable of propagating and modulating potential electrical variability (see next section).

[9]This is not to say that neurons only establish connections among themselves. In fact, neurons connect with practically all other types of cell in the body.

As we shall see in next section, the capacity of the nervous system for enabling flexible, fast and efficient movement lies in its relative dynamical decoupling from morphological and, in general, metabolic-constructive processes.

7.2 The Dynamical Decoupling of the Nervous System

Through early metazoan evolution, the nervous system emerged as a network capable of producing recurrent and specific dynamic patterns quite independently from the underlying metabolic transformations undergone by the organism. Unlike chemical signals circulating within the organism, which directly interact with metabolic processes due to their diffusive nature, electrochemical interactions between neurons enable recurrent interactions within the nervous system itself. The nervous system constitutes a cellular infrastructure in which metabolic energy is converted into more flexible electrodynamic processes, thus creating a new dynamic level that can be said to be, at relatively short time scales, less subject to the constitutive closure of the organism and, in this sense, able to develop distinctive network topologies and intercellular signalling.

As argued elsewhere (Moreno and Lasa 2003; Moreno and Etxeberria 2005; Barandiaran and Moreno 2006; Barandiaran 2008), the relative *dynamical decoupling* of the nervous system means that the metabolism generates and sustains a dynamical system, while at the same time minimising its functional interactions with it. The metabolic organisation produces and maintains the architecture of the nervous system by providing the energy required to feed its dynamics. Yet, the dynamical decoupling means both that (a) neurons minimise interference in their local metabolic processes thanks to their ion-channelling capacities and (b) the constitutive organisation of the organism (what we labelled in Chap. 1 as its "metabolism") *underdetermines* the activity of the nervous system, which rather depends on its internal dynamics and its embodied sensorimotor couplings with the environment. In a word, the biophysical specificity, high connectivity, embodiment, and situatedness of neural electrochemical dynamics make them largely independent (at least at the time scales relevant for describing functional sensorimotor interactions) from the underlying metabolic organisation.

The dynamical decoupling of the nervous system enables the emergence of several new functions and, more generally, has many consequences. In particular, the recurrent interactions between neurons may give rise to higher-level patterns (such as synchronisations at different temporal and spatial scales) endowed with a higher degree of dynamic complexity. Furthermore, these higher-level patterns include internal selection processes taking place at frequencies that are much higher than those found in any other of the organism's control processes. As a result, no other intercellular system even comes close to having the nervous system's capacity to functionally correlate so many elements and, at the same time, to selectively modify their states so quickly. In this respect, the specificity of the nervous system is therefore its ability to generate an enormous variety of states (configurations) per unit of time, and to coordinate an immense number of state transformations simultaneously.

The decoupling of neural processes raises the question of how to characterise their functional organisation. The active electrochemical conductivity of the neurons, organised in spikes or action potentials, allows for a stable combination of them that, added to the network structure of the nervous system and the action of neural modulators, generates a domain that is highly dimensional, non-linear, recurrent, and recursive.[10] Nonlinearity allows distinctiveness of states, while recurrence, provided by the structure of the network, allows circularity and re-entry (Edelman 1987). Recursivity, on the other hand, occurs because spikes can affect themselves through the neural modulators they activate.[11] As a result, the effective dimensionality of the system is constantly being redefined by its own activity (for details, see Barandiaran 2008).

It is worth stressing that neural dynamics are not only sustained by but also causally connected to dynamics at other organisational levels in the organism. For instance, neural patterns functionally modulate metabolic processes in the muscles. Yet this causal connection is largely independent from energy and material aspects, because it is established through the *patterns* of spikes and not through the energetic properties of these very patterns. For example, the motor action triggered by neural spikes is not determined by the electrochemical energy that constitutes action potentials, but rather by their configuration, which selects metabolic energy to produce movement. By virtue of their sequence of changes in amplitude and frequency, the neurotransmitters that neurons generate (when a given pattern of spikes arrives from other neurons) trigger a cascade of chemical processes in the muscles, which convert patterns of spikes into mechanical work.

From the point of view of the organism's overall organisation, the nervous system plays a very complex and manifold functional role. As mentioned, it is subject to closure at a larger time scale than that at which its distinctive dynamics occur. At the same time, the nervous system exerts a higher-level control over musculoskeletal dynamics, independent from the particular energy-related details of how movement is achieved. In addition, the nervous system monitors many intercellular processes through the so-called neuroendocrine systems. Even more importantly, the nervous system is functionally connected to the external world

[10]It is commonly accepted that the primary operational primitives of the nervous system are the changes of neuron membrane action potentials over time (generally in the form of spikes), which conserve dynamic variability in terms of spike frequencies and time distance between spikes. Synaptic connections, on the other hand, specify a connectivity matrix (the transformation functions between primary operational primitives). Actually, the basic connectivity matrix is modulated by neural modulators (local and global synaptic modulators and action potential threshold modulators), which operate at a slower speed (neural modulators are secondary operational primitives because they become operational primitives in virtue of their effect on the spikes). These primitives (spike rates, inter-spike intervals, time of arrival, gas-net modulation, synaptic modulators, axonal growth, etc.) constitute the neural domain.

[11]In fact, neural dynamics depend not only on inter-neural relations. In the neural system, indeed, there are many other non-neuronal cells (glia, astrocytes . . .) that also influence the inter-neuronal processes. It seems that these influence is much more important in vertebrate's neural system (Bullock et al. 2005).

through sensorimotor interactions. As a result, neural dynamics have a very specific functional status in the organisation of autonomous systems.

The specific characteristics of the nervous system have led Maturana and Varela to claim that it can be legitimately said to realise what they call "operational closure" (Maturana and Varela 1987). In their words:

> as a network of active components in which every change of relations of activity leads to further changes of relations of activity. Some of these relationships remain invariant through perturbation both due to the nervous system's own dynamics and due to the interactions of the organism it integrates. (p. 164)

According to these authors, these properties even justify characterising the nervous system as realising itself an *autonomous* organisation. Now, the fact that neural dynamics, although decoupled, are also embodied in a complex organisation which might itself be, as we claimed in Chap. 6, a higher-level autonomous system (i.e. the multicellular organism) makes the issue of its closure (or autonomy) very intricate. For example, would the nervous system realise functional closure and autonomy *in the same sense* than first-level and second-level organisms? In particular, does it define its own specific norms, or is it subject to the normativity of the encompassing biological organisms?

To provide a relevant answer to these questions, one should first understand in more precise terms in what sense the nervous system can consistently said to be at the same time decoupled *and* embodied; in turn, a better understanding of the embodiment of neural dynamics can be had by looking at the way they have changed throughout the evolution of certain animals. Once the question of the interplay between embodiment and decoupling is addressed, we will come back (Sect. 7.5) to the crucial conceptual issue concerning the organisational status of neural dynamics.

7.3 The Evolution of the Nervous System[12]

Although during the earlier stages of metazoan evolution, the nervous system was selected because it allowed fast, plastic, and more efficient movements, we have already emphasised that what matters most for our purposes is its potential for supporting the further complexification of animal behaviour.

Indeed, even the behaviour of primitive metazoa endowed with the most primitive nervous systems, such as cnidaria, may be much more sophisticated than that of any other unicellular or multicellular organism lacking a nervous system. In fact, the appearance of the nervous system (along with other correlated innovations, as the musculoskeletal system) has promoted the appearance of new kinds of multicellular organisms, capable of a huge variety of behaviours. In addition, many motile parts of the organism can be led to move synchronically when needed or, conversely,

---

[12]This section is based on Moreno and Lasa (2003) and Moreno and Etxeberria (2005).

decoupled; strength can be modulated and applied to selected targets. Features of the environment can be categorised, processed, and functionally integrated by sensor organs. Last, but not least, the nervous system has permitted new ways of fast and targeted regulation of the multicellular internal organisation.

In this section, we focus on the evolution of the nervous system, and suggest that the complexification of animal behaviour should be understood in the light of the interplay between the embodiment and decoupling of the nervous system. Each new stage in this process goes with the establishment of new relations with the organism's body (in particular, of vertebrates) as well as new specific neural dynamics and architectures (Fig. 7.1).

7.3.1 Body Plans and the Complexification of the Nervous System

The background assumption of this chapter is that the evolution of animal agency is closely linked to the evolution of the nervous system, whose dynamics are intrinsically intertwined with a complex biological organisation (Moreno and Lasa 2003). Consequently, the potentialities of the nervous system cannot evolve independently from changes in the general organisation of the body (and specifically, the "body plan"[13]). Indeed, as Chiel and Beer (1997) have pointed out, the appearance and development of more complex kinds of behaviour is the result of reciprocal interactions occurring – and constraints exerted – between the nervous system, the body, and the environment. As mentioned, the nervous system participates in the functioning of the metabolism through the neuroendocrine system.[14] In turn, metabolism ensures the adequate maintenance of the nervous system (construction, repair, and adequate energy supplies).

Since relatively simple forms of nervous systems seem to be sufficient for allowing complex and diversified behaviour patterns, would the appearance of more sophisticated neural networks result in some sort of evolutionary benefit? And why has this development occurred mainly in one particular evolutionary line (that of vertebrates)?

As for the first question, selection operates in accordance with the functionalities at work, not in accordance with (possible) future advantages. And, at the beginning,

---

[13]A body plan is the set of constraints harnessing the development of the structural features of a whole phylum of multicellular organisms. It is the framework that guides the way its body is laid out. Once fixed, a body plan becomes a constraining (in the sense of either "limiting" or "enabling") factor in the evolution of a given line of animals, since adaptations only take place inside the architectural limits of the ancestral body plan (Hickman et al. 2001).

[14]In comparison with the nervous system, the functioning of the neuroendocrine system is slower and more durable. As we will see in the next section, in certain animals, in addition to the neuroendocrine system, there is also a direct takeover by the nervous system of some body functions.

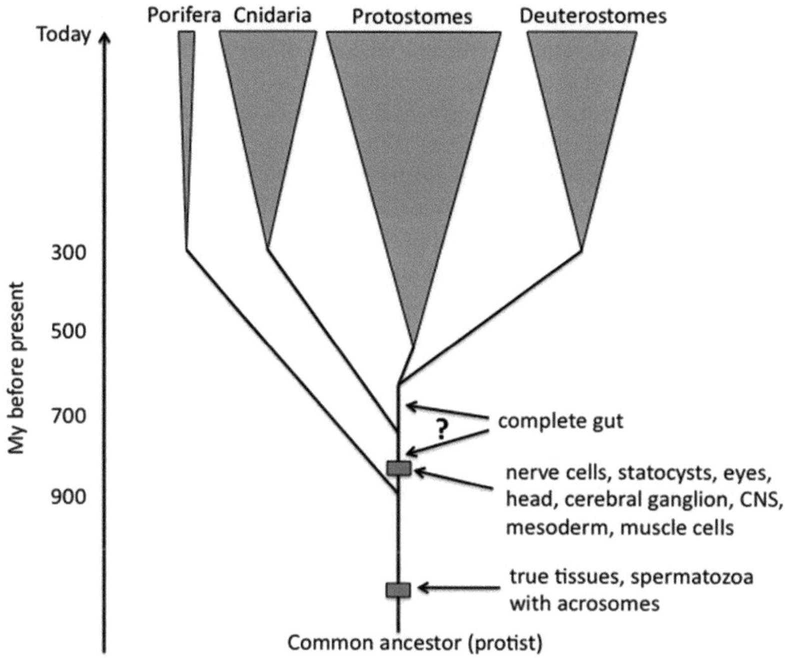

Fig. 7.1 Scheme of the evolution of the nervous system in the different phyla (*Credits: Juli Peretó*)

most innovative changes did not seem to convey adaptive advantages, just viability for differently organised animals. However, if in the process of exploring and successfully occupying new empty niches created by larger animals some were capable of more efficient and flexible motility, and if this capacity was a consequence of the specificity of their nervous system, then this type of nervous system would have conferred a selective advantage. Therefore, what was actually a selective advantage was the capacity to occupy a new niche. As we shall see shortly, this was precisely the case with vertebrates' nervous system.

As for the second question, reasons also exist which might explain why the emergence of a larger, more complex nervous system took place in vertebrates. Although some invertebrates possess relatively large nervous systems (for example, some cephalopods have big brains, while the brain of certain octopuses can contain more than 250 million of neurons), vertebrates' body plan has allowed a different form of embodiment that, in turn, enabled the evolution towards more complex nervous dynamics and in particular, more centralised neural architecture (encephalisation).[15] Vertebrates' encephalisation, in turn, requires an adequate evolutionary explanation.

[15]In fact, the big nervous system of certain cephalopods is much more evenly distributed than in vertebrates: for example, more than half of the neurons of the big octopus' nervous system are located in the arms themselves.

It might be argued, for instance, that the evolution towards a more complex nervous system in vertebrates is related to the colonisation of a terrestrial environment, since an aquatic environment seems less favourable to the evolution of complex agential capacities than a terrestrial one (terrestrial life faces a far more stressful range of environments than marine life, see Raff 1996). This hypothesis is congruent with the fact that certain reptiles, such as crocodiles, also show a primary form of neocortex. In this respect, the tendency of vertebrates to develop larger and more complex nervous systems can only be assessed in the long run.

We will discuss next in more detail how the evolution of the nervous system has become so specific in vertebrates, whose body plans enabled the further emergence of new and more complex agential capacities.

7.3.2 Towards Further Decoupling: Autonomic and Sensorimotor Nervous Systems

The progressive increase in size of multicellular organisms during evolution posed serious problems for the deployment of versatile and strong motility. When animals grow bigger in size, control of fast, precise movements becomes more difficult for at least two reasons. First, the fast, strong movement (especially in terrestrial environments) of large body masses requires some kind of surface for muscles to be inserted in. In invertebrates, this is accomplished, to some extent, through the external skeleton. For larger body masses, however, it would have been too heavy and body growth would also be constrained by this external rigid cover (Storer et al. 1979). Secondly, whereas small animals do not require special systems for distributing nutrients and oxygen or for collecting their residual substances from catabolism (as all their cells are close to nutrient sources and the environment) larger animals need more complex circulatory systems (Nilsson and Holmgren 1994). For these reasons, while their bodies and nervous systems are equally successful when competing in small niches, in the competition to occupy the niche of large-size multicellular organisms, invertebrates were eliminated by vertebrates.[16]

The appearance of vertebrates about 525 mya (i.e. during the Cambrian radiation) set the conditions for the reorganisation of the relationship between the body and the nervous system. One of the key features of this reorganisation is the functional differentiation of the nervous system into two different (sub)systems that, although functionally integrated (because of closure), perform different tasks: the Autonomic

[16] Lacking fine-tuned control of blood circulation, large invertebrates (pogonophores, giant cephalopods, and others now extinct) tire easily and their vascular system is forced to work close to its physiological limits (Abbott 1995). As we have emphasised, like other large invertebrates, large octopuses do not display motility that is as efficient as that of vertebrates. They rely on a system involving three hearts and permanently high blood pressure.

Nervous System (ANS), devoted to the control of viscera,[17] and the SensoriMotor Nervous System (SMNS), devoted to the control of interactive processes.

In vertebrates, an important part of neural resources is devoted to controlling the metabolism through the ANS, (through direct neural modulation of the functioning of different viscera, such as the circulatory and respiratory systems).[18] In particular, the ANS can modify the pressure and flow of nourishing blood in different body areas and organs by means of direct neural control, contracting or dilating the vessel wall or adjusting pump (heart) functioning (Sherwood 1997). The cardio-circulatory system, as mentioned, allows muscles to work in a quicker and more efficient way, which improves animal movements: as a consequence, the ANS allows the muscular system to mobilise a large body mass with speed and strength. This also implies a better control (through the highly efficient circulation of hormones, peptides, and other regulatory substances) of the metabolism of other internal organs (viscera), as well as a better modulation of different organs and their functions: digestion, respiration, sexual activity, etc.... (Moreno and Lasa 2003). Reciprocally, the development of this kind of circulatory system provided the adequate energetic maintenance of big neural concentrations (Levi-Montalcini 1999) and, in particular, of encephalisation.[19] There is, therefore, a mutual relationship between the evolution of the nervous system and the changes in body organisation. Namely, the complexification of the former has induced the complexification of the latter and vice-versa.

The functional differentiation of the nervous system entails many other fundamental consequences for the evolution towards cognition. As discussed, the ANS receives information from all the viscera, integrates it independently from the rest of the nervous system, and sends signals back to the viscera so as to maintain adequate homeostasis (Kandel 1995). In turn, the SMNS (the "system of the exterior") has become increasingly specialised in controlling sensorimotor activity quite independently of metabolism coordination.[20]

[17]During evolution, the ANS has been associated to other neural structures (like the limbic system), all of them performing fine-tuned control over body functions. Together, these structures constitute what the neurobiologist Gerald Edelman (1989) calls the "Nervous System of the Interior".

[18]As mentioned, the basic way for the nervous system to control the functioning of the body is through the neuroendocrine system, which operates through highly specific substances (hormones) distributed by the circulatory system. Unlike the neuroendocrine system, which is based on diffusion and is largely distributed, a vertebrate's ANS is a centralised system, which operates mainly through fast, direct, and targeted neurally-channelled control.

[19]The concentration of neurons – for example, in ganglia – is associated to the realisation of tasks that require a certain degree of complexity. As we have said, invertebrates' nervous systems are in general much more distributed than vertebrates', where we find an evolutionary tendency to concentrate neurons in the head.

[20]Alongside the SMNS, there are the structures that coordinate the sensorimotor tasks performed by the SMNS and the internal organ control tasks performed by the ANS. In particular, coinciding with the appearance of reptiles (about 310 mya) another specific structure appeared in the brain, the limbic system, which is a system of interconnected nuclei that bridge the ANS and the SMNS (Gloor 1997). The limbic system organises the flow from the ANS to the SMNS of both

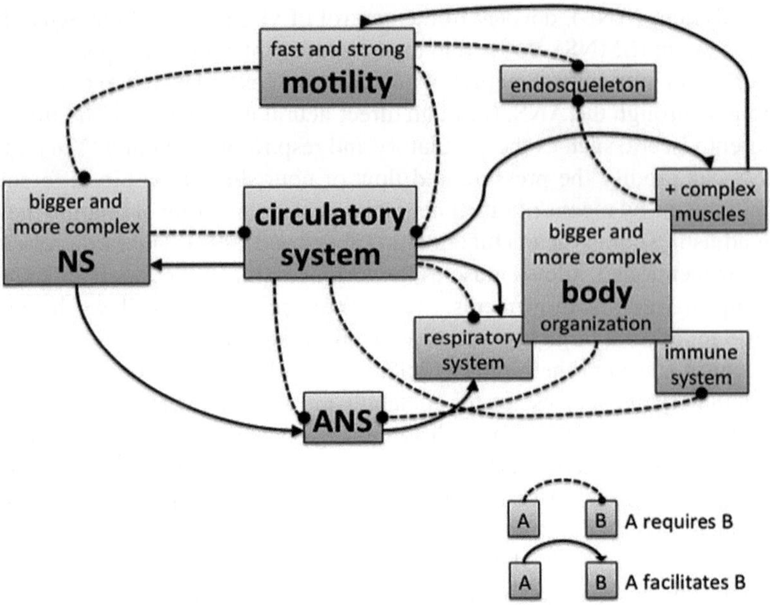

Fig. 7.2 Scheme of the role of the ANS within the nervous system of vertebrates (Adapted from Barandiaran (2008) by Juli Peretó)

The main point here is that the two subsystems ANS and SMNS perform functionally distinct roles, which in turn relies on the fact that (1) they possess a different degree of connectivity and dynamic complexity within and between them, with this connectivity and complexity being higher in the former than in the latter case, and (2) their local dynamics obey different rules (for instance, the neurons of the ANS have the capacity for spontaneous polarisation and depolarisation) (Fig. 7.2).

From the perspective of the evolution of agency, the functional differentiation within the nervous system is what sets the conditions for conciliating its embodiment in the metabolism and the further increase of its dynamical decoupling. Indeed, since the nervous system controls both metabolism and movement, and since as both tasks tended to increase in complexity, the activity of the nervous system would otherwise become less and less efficient and increasingly unreliable. Instead, the differentiation between the ANS and SMNS, and the resulting "division of labour", permitted a substantial reduction in this burden, enabling in particular the exploration of increasingly complex and efficient sensorimotor dynamics.

---

neural connectivity and the secretion of peptides (as well as other neuromodulator substances) that can modify qualitative aspects, such as speed, in the operation of many brain circuits. The limbic system organises the flow back from the SMNS to the ANS as well, by means of neural connectivity.

As we will see next, these massive changes paved the way towards a process of further reorganisation of the relations between the body and the nervous system, leading to the appearance of new, even more complex forms of agency: emotions and a primary form of awareness.

7.4 The Appearance of Consciousness

The differentiation between the ANS and SMNS subsystems requires the establishment of new forms of *coordination* between them, which includes complex interactions with the environment, as well as with visceral and metabolic states. Research into this coordination has shown that in fact the modulatory capacity of the ANS is recruited to adaptively control the SMNS activity (Bechara 2004; Damasio 1994). According to the latter author in particular, such control is the biological basis of emotions.[21] From this perspective, therefore, emotions are seen as a fundamental part of the functional organisation of the nervous system (see Sect. 7.5. below). Emotional processes happen not only at moments of stress when the organism is at a high degree of arousal but at any time.

What matters most for our purposes here is that, during the evolution of terrestrial vertebrates, the coordination between the ANS and SMNS is performed through new structures having appeared in the brain. These new structures (see below for details) – typically located in the cortex – are endowed with an increasing capacity for functional integration, and set the basis of the realisation of a higher-level organisational closure, which includes the more distributed dynamic organisation of the ANS and SMNS (supported by the evolutionary older parts of the vertebrate NS, like the brain stem). In particular, this higher-level organisational closure will be endowed with new capacities of higher-level control exerted on neural dynamics, as well as on intercellular metabolic organisation.

As Christensen (2007) has pointed out, such a hierarchical architecture of the nervous system enhances both control and flexibility because adaptive contingencies can be very complex and can change dramatically. The higher-level organisation should be able to form and reform goals for action based on shifts in any of a large range of agent-based and environment-based factors. The lack of higher-level control over neural dynamics would result in stereotypic forms of motor patterns such as basic walking movement; whilst its presence allows adjusting the action to the circumstances (e.g. brain stem postural control) and set goals such as direction and speed (determined by the cortex).[22] As mentioned in Sect. 7.1. above, the

[21]Many neuroscientists have defended a similar view to Damasio's (see for instance Ledoux 1996; Lewis 2005), postulating that emotions arise as the result of the complex interplay between the ANS and the SMNS and the functioning of the viscera.

[22]As Christensen argues, this hierarchical scheme is supported by classical experiments involving the sectioning of the central nervous systems of cats (Brown 1911; Sherrington 1947). When the

appearance of emotion-based control in the organisation of the nervous system seems to be consistent with the claim put forward by Christensen according to which what is under selection when agency increases in complexity is the set of factors that account for the evolution of high-order control.

Let us examine in more detail how this could happen. The interactions between metabolic states, the autonomic, and the sensorimotor nervous system are the basis of what Edelman (1992) calls "primary consciousness",[23] and Damasio (1994, 1999) "nuclear consciousness". According to this second author, being aware of something would be the process of linking the sense of "self" to a given stimulus or action. In other words, an animal is conscious as soon as it is aware that its actions and perceptions are related to its own body. Thus, the animal needs continuous feedback from viscera and other homeostatic detectors in order to have a sense of self, i.e. to be aware of the state of its own body, potential dangers, or its state of pleasure or needs at any given moment. Awareness involves a bidirectional link, going towards the viscera and in general to the metabolic side of the body through emotional phenomena and, at the same time, to the environment through sensorimotor coordination.

Edelman has developed a detailed neurophysiological approach to the emergence of primary consciousness. According to him, primary consciousness emerges when large "neural assemblies" are formed, and it requires a high degree of functional integration. In particular, the appearance of primary consciousness depends on the evolution of three functions:

1. The development of the cortical system in such a way that when conceptual functions appeared they could be strongly linked to the limbic system, thus extending already existing capacities to carry out learning.
2. The development of a new kind of memory based on this linkage, which performs this task in accordance with the demands of limbic-brain stem "value systems".

---

brain stem and spinal cord are isolated from the forebrain, a cat is still able to breathe, swallow, stand, walk, and run. However, the movements are produced in a highly stereotyped, robotic fashion. The animal is not goal-directed, nor does it respond to the environment. Thus, the brain stem and spinal cord are responsible for producing basic movement coordination, but not higher-level environmental sensitivity or goal-directedness. Instead, a cat with intact basal ganglia and hypothalamus, but a disconnected cortex, will move around spontaneously and avoid obstacles. The animal can perform relatively complex tasks such as eating and drinking and can also display emotions. Evidently, this level of motor control is responsible for the core elements of motivated behaviour. The hypothalamus plays an especially prominent role in integrating the activity of the autonomic and somatic motor systems. But the experiment shows that the most complex forms of behaviour involve the cortex. This area of the brain performs "episodic control", adjusting goal-directed action in relation to local contingencies. And for that, what is needed is not only highly integrated perceptions, but also perceptions that are associated with the animal's goals, values, and environmental context.

[23]Edelman uses the term "primary consciousness" to refer to the varieties of perceptual awareness that humans share with many animals, thanks to which they are able to integrate observed events into memory so as to create a subjectively "aware" perception of the present and immediate past of the world around them.

This "value-category" memory allows conceptual responses to occur in terms of the mutual interactions of the thalamocortical and limbic-brain stem systems.
3. A special circuit that allows for continual re-entrant signalling between the value-category memory and the on-going global mappings that are concerned with perceptual categorisation in real time.

Two distinct structures of the nervous system play a crucial role in the emergence of consciousness: the limbic and the thalamocortical system. The first is fundamentally related with the regulation of the body, since it controls all information relayed from the body to the brain (and vice versa), including control of emotions. The second controls mainly sensorimotor tasks. The thalamus connects a variety of subcortical areas and the cerebral cortex, and its functions are related to the control of the sensory systems (except for the olfactory system), such as the auditory, somatic, visceral, gustatory, and visual systems. As mentioned above, the joint action of these two brain structures coordinate (by exerting a higher-level control over) the relations between the ANS and SMNS, allowing increasingly inclusive forms of functional integration of many local neural dynamics.

How do these two systems actually achieve such global functional integration? A key concept is what Edelman calls "re-entrant signalling", which *synchronises* these different neural ensembles. Their global functional integration results in the formation of a "coherent perceptual scene" associated with emotional states. This is the basis of the appearance of a world of "meaningful perceptions" (see also the following section). Within the thalamocortical system, local perceptions are bound into associated bundles in a functionally integrated way. Since what Edelman calls "the demands of the value systems of the individual animal" clearly are connected to emotions, this also implies that the construction of the scene is linked to the body and emotional states. The integration of all these functions therefore requires a cross-correlated integration with a huge number of patterns of neural activity, both distributed and localised (Fig. 7.3).

The process leading to the integration of distributed neuronal activity has been proposed by Edelman and Tononi as the indicative "mark" of the emergence of primary consciousness, insofar as it seems to be clearly correlated with being awake and disappears in situations of deep sleep, anaesthesia, and epileptic episodes (Tononi and Edelman 1998; Edelman and Tononi 2000).[24] More specifically, Edelman and Tononi argue that the formation of a complex dominant neurodynamical structure (what they call a "dynamic core") would be the basis of the unity of conscious experience and of its relatively short duration (it is only perceived over a few hundred milliseconds, a period corresponding to a few gamma cycles).

The dynamic core therefore emerges as soon as the neurodynamic organisation is subject to a set of high-level control mechanisms, which in turn require the

[24]This kind of neurodynamic organisation in thalamocortical areas occurs in the gamma frequency band (Llinás et al. 1998). This has led some authors to suggest that timing in this frequency band in the visual areas may be the correlate of conscious visual experience (Crick and Koch 1990) or of perceptual consciousness (Engel and Singer 2001).

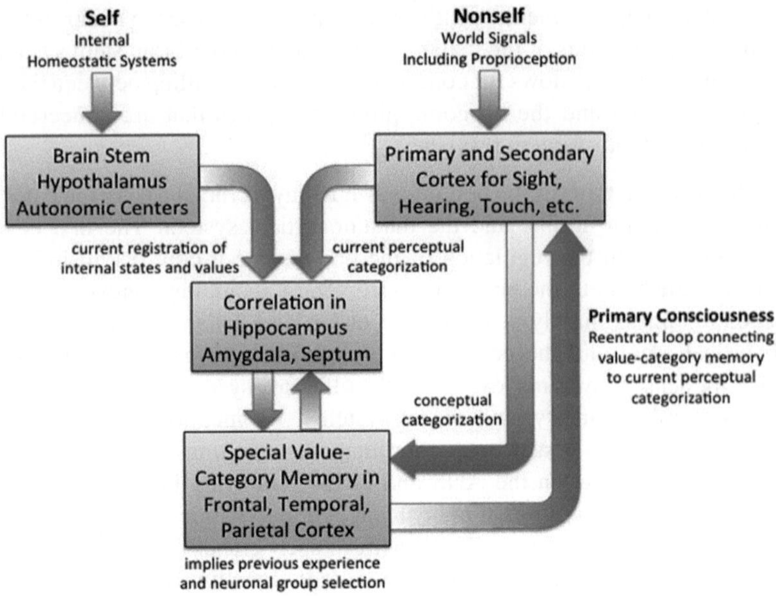

Fig. 7.3 Structure of the feedback relations between different parts of the brain supporting the emergence of primary consciousness (Adapted from Edelman (1987) by Juli Peretó)

integration of perceptual and emotional processes. The intertwined relationship between perceptive and emotional processes that together realise, we submit, a higher-level closed causal organisation, has also been emphasised by Lewis (2005), Thompson (2007), and Pessoa (2008). In next section we will explain in what sense this neurodynamical organisation plays a crucial role in the emergence of cognition.

The (succinct) account of conscious control described in these pages might be put to work when one faces the problem of determining which animals are capable of conscious agency. In particular, it can guide the search for what Seth (2009) has called "explanatory correlates of consciousness", namely, neural organisations that have been correlated with conscious activity in humans by exhibiting both integration and differentiation. In this respect, there are indeed many indirect arguments in support of the claim that higher vertebrates, such as most mammals and probably certain birds, are capable of conscious behaviour. In turn, the case of highly evolved invertebrates, such as octopuses, seems different to the extent that in these animals the nervous system (despite the discovery of very interesting behaviours) is more distributed and less integrated. As David Edelman and Seth (2009: 482) have pointed out:

> radical differences between cephalopod nervous systems and those of vertebrates are exemplified by the parallel, distributed architecture of the octopus loco motor system.

The example of cephalopods shows that the evolution of the nervous system and agency may also attain considerable complexity by following a very different trend.

What matters, as Godfrey-Smith (2010, 2013) has pointed out, is that the nervous system of an octopus is less centralised than ours (see also Zullo et al. 2009). In fact, more than half of an octopus's neurons are located not in the animal's central brain at all, but in its eight arms. As he writes:

> it is as if each arm has a mind of its own. Or perhaps in an octopus we see *intelligence* without a unified *self*[25]

In short, conscious agency is the result of a specific evolutionary pathway in which agency has become more and more complex within the framework of a specific set of enabling constraints in the vertebrate body plan. This evolutionary process should be understood in terms of a complex balance, found between the embodiment and the increasing decoupling of the neural organisation with respect to the body, as well as the control exerted, through emotions, over sensorimotor dynamics.

7.5 Cognition and the Emergence of Neurodynamic Autonomy[26]

So far, we have discussed why metazoan multicellular organisms endowed with a nervous system were able to evolve towards higher degrees of behavioural complexity. In particular, we have argued that the development of an increasingly complex nervous system was enabled by a very specific body plan, namely, the vertebrate body plan, which allows fast, plastic, and strong movement at a larger size. This in turn enabled access to new niches and therefore provided an evolutionary advantage to be selected for. At a given stage of this evolutionary process (primary) consciousness appeared, which provided the capacity for a higher degree of hierarchical control over, and integration of, neural dynamics.

As we shall explain next, once this evolutionary stage is reached, the closed organisation of the nervous system itself realises a distinct level of *autonomy* (Barandiaran and Moreno 2006; Barandiaran 2008). What does this mean? In Sect. 7.2., we briefly mentioned that according to Maturana and Varela, the nervous system can be considered a dynamical organisation achieving a form of closure. As Thompson (2007) points out:

> any change of activity in a neuron (or neural assembly) always leads to a change of action in other neurons (either through directly synaptic action or indirectly through intervening physical and chemical elements). Sensory and effector neurons are not an exception because any change in the one leads to changes in the other, such that the network always closes back upon itself, regardless of intervening elements. (p. 50)

[25]http://news.harvard.edu/gazette/story/2010/10/thinking-like-an-octopus/

[26]This section is largely based on Barandiaran and Moreno (2006) and Barandiaran (2008).

By relying on the technical characterisation of closure provided in Chap. 1, we do claim that the nervous system realises a closed organisation that in turn implies that the different structures and mechanisms that modulate the flow of neural spikes might be shown to behave as constraints within the neurodynamical domain. In this chapter, we do not provide a full-fledged justification of this claim that we take as a reasonable working hypothesis. Actually, even in relatively simple nervous systems there seems to be many levels of constraints that functionally act on the neural dynamics. Globally considered, the nervous system controls its perceptual inputs through control cycles of action-perception behaviour (Powers 1973) and can also modify its connectivity matrix through the interaction between primary and secondary primitives (see footnote 8 in Sect. 7.2).

As previously explained, as the complexity of the nervous system increases, this minimal form of neural closure was re-organised, generating new forms of hierarchical control. In the previous section, we described how the nervous system of certain evolved vertebrates is capable of achieving a very high degree of functional integration of large synchronised neural structures. These neurodynamic structures, in turn, are organised as a global closed self-maintaining network, functionally coupled with the external world as well as with the bodily organisation of the animal. In particular, the resulting organisation possesses agential (sensorimotor) capacities, as well as higher-level control functions exerted over these very dynamics, some of these control functions being in fact higher-order regulatory ones. As a consequence, we hold that the resulting neural organisation may be legitimately said to realise a distinct level of autonomy that is embedded in, but distinct from, second-level (multicellular) autonomy.

A crucial aspect of this higher-level neurodynamical autonomy is the existence of a hierarchical control exerted over ongoing sensorimotor activity (i.e., agency). It is worth mentioning that this kind of control, which involves emotions, is an often-neglected aspect of embodiment, more specific than the general, usually more emphasised, metabolic and sensorimotor one (Ziemke 2003). As it has been argued by Barandiaran and Moreno (2006) and more extensively, by Barandiaran (2008), neurodynamic autonomy, consists in a self-regulated closed network of dependencies between neurodynamic structures.[27] On the one hand, this closed network is linked to the sensorimotor couplings and, on the other hand, it monitors the body processes. The neurodynamic organisation must be able to perform these functions simultaneously while maintaining its own coherence by itself, likewise through internal higher-level control.

In our own view this neurodynamic closed organisation satisfies also the criteria of autonomy because, (1) this neurodynamic closed organisation is self-regulated

[27]In the mentioned studies, the concept of dynamic structure is defined as "the subset of internal variables and their relationships involved in a certain sensorimotor coupling. A dynamic structure emerges when (for a given time window) we can systematically reduce the dimensionality of the internal operational organization of the NS to predict the behavior of the system" (Barandiaran and Moreno 2006: 177)

(it selectively modulates the neurodynamic structures); and (2) it is at the same time inextricably linked to the aforementioned conscious agency (the neural constraints governing sensorimotor couplings normatively contribute to the self-maintenance of the autonomous neurodynamic domain). The idea of neurodynamic autonomy advocated here is very similar to that originally put forward by Barandiaran and Moreno. The main difference lies in the fact that we suggest identifying the rather abstract concept of "dynamic structures" with Edelman's more neurologically specific concept of "dynamic core". Moreover, in our account, the role of functional integration and control are of paramount importance.

Now, following Barandiaran and Moreno, we claim that neurodynamic autonomy *is the organisational ground for the cognitive domain.* When neurodynamic autonomy is realised, the self-determination of the neural dynamic organisation becomes the source of the specific teleological and normative dimensions of its constitutive, agential, and regulatory functions of this domain, that is, the "locus" of identity shifts from the metabolic and developmental level of organisation to the neural one (Barandiaran and Moreno 2006; Barandiaran 2008). From the autonomous perspective, cognition refers to the capacity for higher-level neurodynamic control over both sensorimotor and body processes. Cognition involves a world of meaningful (value-charged perceptions) interactions for the animal, and does not consist merely in the capacity of sustaining perceptually guided behaviour. Rather, it inherently involves the successful managing of attention and emotional regulation. Cognitive capacities emerge when higher-order functions specifically constrain sensorimotor couplings, so as to enable sufficiently accurate, meaningful perceptual interactions.

At least in its most minimal sense (that involving perception, memory, and emotion), cognition requires the adequate and complex balance between decoupling, embodiment, and control provided by neurodynamic autonomy. In this respect, the focus of this chapter on the evolutionary stages, which have led to the emergence of cognition, was precisely aimed at putting adequate emphasis on the *qualitative change* in the relations between the neural domain and the multicellular metabolic organisation occurring when neurodynamic autonomy is realised. In turn, as discussed, neurodynamic autonomy consists in a specific way for the multicellular organism, dealing with the establishment of an equilibrium between the fact that neural dynamics are highly decoupled and, yet, deeply embodied in the encompassing metabolism.

7.6 Cognition and Social Interaction

As we pointed out at the end of Chap. 4, organisms establish different relationships amongst themselves, some of which (communicative relations) involve direct signal exchanges. These signals affect ontogenetic metabolic changes, inducing reciprocal collaborative relations such as the formation of different kinds of multicellular organisation. Once organisms with a nervous system appear, they establish a different form of communication, which affects sensorimotor couplings; this occurs, for example, in the case of social insects.[28]

In this sense, communication essentially involves coordinated behaviour between individuals, in order to ensure either their own respective self-maintenance or that of the global entity. Just like in the example of multicellular organisations, these communicative interactions play a fundamental role in giving rise to higher-level, self-maintaining organisations with different degrees of integration. Beehives, for example, are considered one of the most integrated forms of social organisation, in which individual organisms undergo major organisational transformations in order to ensure social maintenance. However, one of the most important differences between these highly integrated social organisations and multicellular entities is that no topological border exists and therefore individuals can (and must) move freely in a wide spatial domain.

Maturana and Varela (1987) define sociality as a form of organisation generated by:

> a particular internal phenomenology, namely, one in which the individual ontogenies of all the participating organisms occur fundamentally as part of the network of co-ontogenies that they bring about in constituting third-order unities. (p. 193)

More recently, and within the autonomous perspective, various authors (De Jaegher and Di Paolo 2007; Fuchs and De Jaegher 2009; McGann and De Jaegher 2009; Di Paolo and De Jaegher 2012) have emphasised the importance of interactive experience for both the development and current functioning of cognitive capacities, claiming that interactive elements shape and may even constitute socio-cognitive mechanisms. Although these authors focus mainly on human social interactions, they provide a definition of social interaction that is broad enough for our purposes here. Social interaction has been defined as:

> a co-regulated coupling between at least two autonomous agents, where: (1) the co-regulation and the coupling mutually affect each other, constituting an autonomous self-sustaining organisation in the domain of relational dynamics and (2) the autonomy of the agents involved is not destroyed (although its scope can be augmented or reduced). (De Jaegher et al. 2010: 442–443)

We can therefore take these two definitions as our starting point for addressing the relation between social interaction and the origin of cognition.

When we focus on social interactions among cognitive animals, we face another, much more complex form of communication involving emotions and conscious agency. As stated in the previous section, the successful maintenance of neurodynamic closure requires sensorimotor couplings that involve both sufficiently accurate perceptual capacities and higher-order control through emotions. Yet, the

[28]This does not mean that communication between animals is established only through sensorial signals. In fact, in many cases, chemical interactions are also very important. For example, the members of an ant colony share food and other fluids, thus inducing socially coherent patterns in a phenomenon known as Trophallaxis.

maintenance of neurodynamic autonomy *does not depend only on the activity of the individual; it is also a social process*. In particular, in this social process, *affective* interactions are of fundamental importance. It is well known that the ontogenetic development and even adult life of highly evolved vertebrates is crucially dependent on affective interactions. For example, rodents and many birds establish empathic relations (Barthal et al. 2011; Panksepp and Lahvis 2011; Fraser and Bugnyar 2010) and this is also supported by the fact that these animals express facial emotions (Erickson and Schulkin 2003). The external expression of emotions gives rise to the need to perceive, interpret, and react to other organisms' emotions (and therefore behaviours), thus contributing to a new type of communicative and social behaviour (Shepherd 1994). This subsequently constitutes the fundamental basis of nurturing behaviour.

An interesting example for this study is provided by current experiments involving epigenetic activation in rat stress regulators (via the mother-pups).[29] A mother rat that does not adequately regulate her reaction to stress has almost no interaction with her pups and the lack of such interaction (which has an affective component) does not allow the generation of regulatory reactions to stress in the next generation. Highly nurtured rat pups tend to grow up to be calm adults, while rat pups who receive little affective nurturing tend to grow up to be anxious.

External expressions of emotions also play an important role in predatory behaviour, as well as in both competitive and collaborating aspects of social behaviour. At the same time, these kinds of behaviour generate pressure for cognitive complexification. They are closely correlated with (and contribute to) the development of vertebrates' limbic system (Gloor 1997) and further on in evolution, are also linked to a special part of the ANS related to control of facial muscles (cranial or social autonomic nervous system (Porges 1997)), which is very important for nonverbal communication (movement of lips, muzzles, scalp, and external ear flaps, for example). This cranial or social component of the ANS is highly developed in mammals.

In short, the close relationship between affective social interaction and the development of conscious agency may point to an interesting possibility; namely, the understanding of the operational function of phenomenological experience in primary consciousness. If the maintenance of the neurodynamic autonomy depends on the maintenance of the affective social interactions – if when the pups of these animals grow up, the affective and cognitive interaction process modulates the development of their neurodynamical organisation – then the way in which the phenomenological experience arises is also dependent on these interactions. If the feedback of couplings between the partners must be regulated so as to generate a continuous causal interaction between neural mechanisms and subjective experiences ultimately achieving closure, then conscious experience would play an operational role. These brief remarks suggest the interest of what is, today, an open domain of research.

[29] See: http://learn.genetics.utah.edu/content/epigenetics/rats/

7.7 Conclusions

In this chapter, we have focused on the evolution of agency in different kinds of organisms. Complex agency emerged as a result of the selective advantage brought about by the exploration of new niches by those large organisms whose way of life was based on motility. Yet, the action of natural selection toward complex agency was only made possible thanks to the appearance of new, more complex forms of integrated multicellular organisms (metazoa) endowed with new body plans.

A key step in this evolutionary process was the development of the nervous system, because it created a huge internal world of fast, plastic connection patterns which were dynamically decoupled from multicellular metabolic processes, therefore opening up the possibility of supporting rich, complex behaviours. The decoupling of the nervous system goes with its embodiment insofar as it exerts several functions necessary for the maintenance of the general metabolism. The development of more complex forms of behavioural agency emerges in close connection with such embodiment, in which the potentialities and limitations of some basic body plans proved to be enabling or requisite for further evolutionary development of new interactive capacities, while those of others hit apparent ceilings.

Within the vertebrate body plan in particular, selective pressures led to appearance of a kind of nervous system not only capable of supporting larger organisms that were both versatile and rapid, but at the same time allowing a self-sustaining process of integrated neurodynamic organisation, thus resulting in the emergence of a new form of agency involving emotions and awareness.

In this process of complexification, the nervous system itself became capable of controlling not only behavioural functions but also the whole body organisation. Now, for the same reason that a living cell is an autonomous system, a nervous system that generates a set of regulatory controls that drives it and maintains its far-from-equilibrium identity should also be considered an autonomous system in the neural domain. This new form of neurodynamic self-maintenance and autonomy is embedded in yet different from the biological one. Therefore, an autonomous level of normativity emerges when the adaptive preservation of the internal organisation of neural dynamics becomes the major source of neurodynamic regulation. And this specific normativity, which governs the whole organism, is the basis of a new form of agency.

As explained earlier, this new form of agency implies a high-level, integrative form of control in the neurodynamic domain: the "conscious mind". Conscious agency is the expression of a new form of autonomy because the structure of the regulatory controls that govern the behaviour of the animal is itself dependent on the maintenance of the functional sensorimotor actions triggered by them. Here again we see that "the being" (in this case, the conscious mind) is ultimately dependent on its "doing" (conscious, higher-level environmental sensitivity and goal-directed sensorimotor behaviour). But the maintenance of coherency between "being" and "doing" requires that the actions fulfil (species-dependent) epistemic norms. In other words, if what the animal does is not congruent with its cognitive expectancies, and if this incongruence cannot be corrected in time, then the whole structure of its regulatory controls would begin to disintegrate, and ultimately so would its own conscious mind. And conversely, the more successful the high-level sensorimotor behaviour of the animal, the more its conscious identity is reinforced.

Thus, in our approach, life is a necessary condition for the appearance of cognition, but only certain living entities are cognisers. Autonomy, of course, is what enables living systems to become cognitive agents. But only a very complex and nested form of autonomous organisation, harbouring in turn a neural autonomous organisation, dynamically decoupled from metabolic multicellular organisation, can support cognition. We therefore disagree with the claim that it is the very organisational properties of living organisms make them cognisers (Varela 1997; Thompson 2007). Although strongly embodied and autonomous, cognition in our view is an emergent phenomenon that can by no means be identified with generic biological autonomy, and which appeared when the nervous system became an autonomous organisation capable of governing the (second order) autonomous metabolic organisation of certain multicellular organisms.

8
Opening Conclusions

In these pages we have explored a vast range of phenomena, spanning from the physicochemical to the human realms, trying to understand life at the intersection between these widely heterogeneous domains.

Since the middle of the past Century, scientific knowledge in biology has increased exponentially. Yet the growth has been much faster in the empirical domain than in the theoretical: we have discovered many biological phenomena but, to date, we have not integrated them into unified frameworks. A fundamental reason for the present situation is that biological sciences are providing us a picture of the biological domain as a universe of extreme complexity. As a consequence, current biological knowledge is disseminated in a number of different domains, with not a few reciprocal inconsistencies; it is dynamical, rapidly changing, and in many cases, at the precipice between old ideas, increasingly criticized, and new ideas, not yet well articulated (though often promising). For these (and other) reasons, current biology is replete with conceptual puzzles and cannot globally evaluate all its own implications.

Here is where the work of both theoretical biologists and philosophers begins. As philosophers of biology in particular, we conceive our investigations as a contribution to the elucidation of the principles underlying this complexity. We undertake these projects not because we believe that ultimately there is a "simplified picture" of this intricate network of relations, levels, and hierarchies; quite the contrary, this complexity is presumably irreducible. Rather it is because we hope that, somehow, we can understand it by discovering hidden and yet encompassing properties and features in the huge amount of complicated experimental evidence that biological sciences provides us with. Admittedly, this is an extremely difficult task, in no small part because there are not pre-established methods and because of the novelty and complexity of the questions. It is therefore a high-risk task, intellectually speaking, but at the same time, a very relevant one for the development of scientific research. This is how, in our view, philosophy of science should be evaluated. A relevant philosophical effort should help overcome theoretical crises;

A. Moreno, M. Mossio, *Biological Autonomy*, History, Philosophy and Theory of the Life Sciences 12, DOI 10.1007/978-94-017-9837-2_8

it should make qualitative progresses when science faces critical challenges and bottlenecks; it should help to connect scientific domains; it should make explicit and critically analyse implicit assumptions.

In this respect, the approach developed in this book lies between philosophy and theoretical biology. It deals with philosophical questions like the nature of autonomy, agency, and cognition, as well as their relations to concepts such as function, norms, teleology, and others; yet, it addresses these questions with close connections to, or even deeply entangled with, current scientific research. It proposes a theoretical framework that can contribute to the integration of different biological domains by favouring a shift of the experimental priorities towards organisational issues. In this sense, it can set up a research program that, in the medium term, can be submitted to empirical testability.

The framework that we have proposed is centred on the idea of autonomy. Our main claim is that the distinctive feature of biological organisms, which distinguishes them from any other natural system, is their autonomy. Biological systems are autonomous systems, which means that, in the most general sense, they contribute to the generation and maintenance of the conditions of their own existence. Autonomy realises a circular relation between "doing" and "being": not only are biological systems able of acting on the world but reciprocally, their activity plays a crucial role in determining what they are. In turn, the circular (teleological) relation between the existence and activity of the system provides a naturalised ground for normativity: its conditions of existence are the norms governing its activity. That is why biological organisms are literally *auto-nomous*, they generate by themselves – at least in part – the norms that they are supposed to follow.

In the past, many authors have expressed views that were closely related, or even coincident, with the general picture summarised above. As repeatedly mentioned in the book, there is indeed a tradition in biology (or more precisely, in theoretical biology, and somewhat in cognitive science) that has promoted an understanding of biological phenomena by appealing to concepts as autonomy, circularity, self-organisation, and related notions. With respect to this tradition, we attempted in this book to make a step beyond this fragmented collection of (more or less) related approaches and models. We have proposed an integrated theoretical framework, on the basis of which the autonomous perspective could be put to work in order to address, in a coherent way, the study of biological phenomena.

In these concluding pages, it might be useful to sum up the main ideas that we have been developing throughout the book by providing a synthetic overview of the autonomous perspective as we conceive it. Also, because most work still remains to be done, we will mention some central issues that the theoretical framework outlined here should handle in (the near) future, on its way toward a comprehensive biological theory.

As explained in the Introduction, our general stance consists in suggesting that the principle of biological autonomy must be understood in the light of three characteristic dimensions that are conceptually distinct and yet inherently related. Biological autonomy has a constitutive dimension, which consists in its organisation's capacity of self-determination. Biological organisation determines itself and, through this determination, grounds normativity, teleology, and functionality in a naturalised way. Biological autonomy also has an interactive dimension through which biological systems promote their own maintenance by acting on their environment. Autonomy is not independence: autonomous systems are not monads, they are inherently agents, engaged in a continuous interaction with their surroundings. Lastly, biological autonomy has an historical dimension, which means that it cannot be understood as spontaneity or self-organisation. The complexity of autonomous systems is also the result of historical processes that it additionally contributes to in order to generate.

The main objective of this book consists in suggesting that beyond the general ideas shared by this theoretical tradition, these three dimensions can be spelled out in precise terms as a set of specific hypotheses on the nature of biological organisation. In the first part (Chaps. 1, 2, 3, and 4), we establish the fundamental tenets of the principle of biological autonomy. The second part (Chaps. 5, 6, and 7) develops the idea that biological autonomy unfolds itself historically, which generates, in particular, increasingly complex, entangled, and hierarchical forms of autonomy. As a result of their agency, autonomous systems interact with each other, mutually affecting their respective organisation and constituting higher-level collective autonomous organisations, without losing their autonomy. These higher levels of autonomy can in turn generate new networks of interactions, which can possibly lead to even higher levels of autonomy, and so on. In addition, these entanglements give rise to complex inter-level (although not necessarily "nested") relations, which affect both the lower-level entities and the emergent, higher-level organisations. Thus, the appearance of increasingly complex autonomous systems goes with the appearance of increasingly complex "environments" constituted by the network of constraints exerted by biological entities on each other.

In extreme synthesis, our account has been structured around the following set of claims.

1. Biological self-determination occurs as a closure of constraints, which characterises what biological organisation actually is. Closure constitutes a fundamental principle of order in the biological domain; in spite of the continuous changes that it undergoes, biological organisation maintains closure, albeit possibly realised in different variants. Moreover, actual realisations of biological organisation in real environmental conditions – what we called 'metabolism' – require regulatory functions. Autonomy requires regulated closure.
2. Biological organisation constitutes an emergent regime of causation, because the relatedness among the constituents generates ontological novelty, and therefore distinctive properties and causal powers. Yet the emergent nature of biological organisation does not imply nested inter-level (upward or downward) causation between the whole and its parts, although we cannot (and we do not want) to exclude this possibility.
3. The closed organisation is the naturalised – and thus perfectly admissible in the scientific discourse – grounding of teleology, normativity, and functionality. Indeed, because of closure, the existence of the autonomous system depends

on the effects of its activity, and at the same time, its conditions of existence can be legitimately taken as norms of its own activity. In addition, closure realises the division of labour among the parts, which is the third requirement for functionality. Hence, from the autonomous perspective, "organisation", "closure", and "functionality" are inherently related concepts, all referring to the same emergent causal regime.

4. Autonomy is not independence. Biological organisms, as dissipative systems, can exist insofar as they maintain specific interactions with their surroundings, and an adequate flow of energy and matter. In positive terms: autonomy inherently implies agency, which is realised as a specific subset of functions whose effects are exerted on the boundary conditions of the whole system. Moreover, biological organisms possess most of the time regulatory capacities exerted on their agential functions, which makes them adaptive agents.
5. Autonomy has a historical dimension; it is not just spontaneous self-organisation. Autonomy appears as the result of an entailment of reproductive cycles, starting from self-maintaining chemical systems, which progressively increase their complexity. Reciprocally, the evolution of biological complexity cannot be adequately understood just as the outcome of natural selection, but results from the fundamental interplay between organisation and selection. In this respect, the autonomous perspective renews biological ontology by organising it around units of autonomy instead of units of selection.
6. Collective associations of unicellular organisms, when they attain a sufficient degree of integration, realise higher-level multicellular autonomy, and can therefore be taken as multicellular organisms. In particular, we have emphasised that the relevant degree of integration requires that the system includes a process of development and, in addition, the capacity of exerting regulation on it, so to maintain the very delicate balance between intracellular and intercellular dynamics.
7. Multicellular organisms may generate cognition, which is a qualitatively new kind of autonomy, and not just as a complex form of agency. A naturalised account of cognition requires understanding it as the result of the evolution of both constitutive and interactive complexity towards radically innovative phenomena, such as emotions, consciousness, meaning, and values. In this sense, although *sui generis*, cognition is a genuinely biological dimension.

The set of claims succinctly recapitulated here constitutes an integrated theoretical and philosophical framework, which is explicit enough to be critically analysed and, we hope, prone to be further developed. Let us then mention some open issues with which the autonomous perspective should deal. There is no specific logic in focusing on these particular issues instead of others, and no *a priori* reasons to find these more important than others; they simply point, we think, to relevant and stimulating research directions.

A first research direction concerns the relations between order and disorder, between stability and variability. As we mentioned in the book, the autonomous perspective takes the closed organisation as the fundamental principle of order in biological systems. Closure of constraints is what makes an otherwise near impossible cluster of processes and reactions occur in an ordered way and, moreover, last continuously over generations of individual organisms. Therefore, further investigations should clarify to what extent (and in what respect) the autonomous perspective constitutes a theoretical departure from the mainstream view adopted by molecular biology during the last 30 years, according to which biological order relies on genetic information. This issue is particularly relevant in a moment during which a lively debate exists on how this very notion should be understood, and its place in biological theory (Griffiths 2001; Ruiz Mirazo and Moreno 2006; Levy 2011; Godfrey-Smith 2014). In particular, increasing experimental evidence (as for instance, the observation of stochastic phenomena at the molecular level, in relation to gene expression and molecular interactions, see for instance McAdams and Arkin 1997) is challenging traditional explanations of stability in terms of genetic information (Kupiec and Sonigo 2000).

It seems to us that in this context, the autonomous perspective could make a relevant contribution by inducing a fundamental shift in focus: order and stability derive primarily from organisation, not from genes, although, as discussed, template constraints play a crucial role in the evolution of high biological complexity. Indeed, the relations between the genetic machinery and the whole metabolic organisation (endowed with its multilevel network of regulatory functions) could be seen, from this perspective, as *fundamentally intertwined* (Meyer et al. 2013) because they have evolved together, enabling the appearance of increasingly complex autonomous systems. In any case, the central issue will be to understand how biological organisation succeeds in maintaining an adequate *trade-off* between stability and variability. On the one hand, it must avoid drift and disruption by canalising (also through regulation, as discussed) the variability and, in some cases, even the stochasticity of processes and reactions; on the other hand, it must leave enough room for this very variability to explore functional novelty, which is a condition for adaptivity, the increase of complexity, and in the end, the long-term maintenance of life.

The autonomous perspective should also undertake a thorough investigation of the principles of biological organisation. In this book, we have tried to spell out some of them, as well as their relations. Autonomous systems realise closure of constraints, regulation, (adaptive) agency, and emerge from the mutual interaction between organisation and selection. Although we spelled out these principles in some detail, most of the work is yet to be done; in particular, we do not as of now have an account of the specific functional architecture that a viable biological organisation must realise to meet with the internal and external requirements for its coherence. Many different organisations of constraints can logically meet the conditions for closure and agency, but not all of them are biologically relevant, or even sustainable in the real world. Therefore, we should clarify (both theoretically and experimentally) which functional closed architectures are feasible and relevant. Just to mention a specific example, it might be argued that autonomy requires that closure be "layered" or "cumulative" in the sense that most functions operate on processes and reactions having already been under the constraining action of other functions. In fact, this is what happens in all real biological systems, at least because any function operates on substrates having been entered by the membrane. To what extent does such cumulative architecture constitute a general principle of biological organisation?

In a similar vein, a very stimulating research direction would be one concerning the levels of organisation. As we have claimed, the autonomous perspective conceives the concepts of organisation and closure (and functionality) as inherently related; as a consequence, levels of organisation correspond to levels of closure or even, in some cases, to levels of autonomy. In Chaps. 4 and 6, we have given some preliminary hints on how our framework could deal with the organisation of ecosystems and, in addition, could provide principled criteria to identify multicellular organisms. In Chap. 7, moreover, we have discussed how multicellular autonomy could bring forth a different form of autonomy, which would support the cognitive domain. Yet there is no full-fledged account of the hierarchy of levels and forms of autonomy, and of the entangled relations between them: how many levels of closure and autonomy are there in the biological domain? How exactly do they depend on each other? How could a cell be externally constrained within a more encompassing multicellular organism to which it belongs, and yet maintain its own autonomy? Why are higher-level autonomous systems constituted by parts that, though strongly constrained, remain autonomous themselves? Only future investigations will provide satisfactory answers to these fundamental questions.

One final, important issue – implicit in our analysis – is that concerning explanatory strategies in biology. As it is well known, in modern biology there are two main explanatory strategies: mechanistic and network theories, which are applied to what Winther (2006) has called "formal" and "compositional" biology. For example, ecology and population genetics are examples of a "formal" type of biology, because they focus on the mathematical relations among abstract entities; in turn, physiology and developmental biology are "compositional", in the sense that they study how systems are constituted by functional parts, and how these parts are assembled into mechanisms. With the increasing incorporation of developmental theories, evolutionary theory ("evo-devo") itself is becoming a synthesis of these different explanatory strategies. In fact, behind (and supporting) each of these methodological and explanatory approaches, there are two different views. On the one side, that which puts emphasis on how biological systems are organised at different levels and domains and on the other, that which focuses on how these different levels and domains are related to each other.

The autonomous perspective aims at merging the epistemological foundations of both explanatory strategies (Moreno et al. 2011): a substantial progress towards their integration would ultimately constitute a relevant step towards a more unified biological science.

References

Abbott, N. (1995). Cerebrovascular organization and dynamics in cephalopods. In N. J. Abbott, R. Williamson, & L. Maddock (Eds.), *Cephalopod neurobiology*. Oxford: Oxford Science Publications.

Adams, F. R. (1979). A goal-state theory of function attributions. *Canadian Journal of Philosophy, 9*, 493–518.

Alexander, S. (1920). *Space, time and deity*. London: Macmillan.

Allen, C., & Bekoff, M. (1995). Biological function, adaptation and natural design. *Philosophy of Science, 62*, 609–622.

Allen, C., Bekoff, M., & Lauder, G. V. (Eds.). (1998). *Nature's purposes*. Cambridge, MA: MIT Press.

Alon, U., Camarena, L., Surette, M. G., Aguera y Arcas, B., Liu, Y., Leiber, S., & Stock, J. B. (1998). Response regulator output in bacterial chemotaxis. *EMBO Journal, 17*(15), 4238–4248.

Amundson, R. (2000). Against normal function. *Studies in History and Philosophy of Biological and Biomedical Sciences, 31*, 33–53.

Anet, F. (2004). The place of metabolism in the origin of life. *Current Opinion in Chemical Biology, 8*, 654–659.

Angerer, R. C., & Angerer, L. M. (2012). *Sea urchin embryo: specification of cell fates*. Chichester: ELS John Wiley & Sons.

Arendt, D. (2008). The evolution of cell types in animals: emerging principles from molecular studies. *Nature Reviews Genetics, 9*(11), 868–882.

Ariew, A. R., Cummins, R., & Perlman, M. (Eds.). (2002). *Functions*. Oxford: Oxford University Press.

Arnellos, A., & Moreno, A. (2012). How functional differentiation originated in prebiotic evolution. *Ludus Vitalis, 37*, 1–23.

Arnellos, A., & Moreno, A. (in press). Multicellular agency: an organizational view. *Biology & Philosophy.*

Arnellos, A., Moreno, A., & Ruiz Mirazo, K. (2014). Organizational requirements for multicellular autonomy: insights from a comparative case study. *Biology & Philosophy, 29*, 851–884.

Artiga, M. (2011). Re-organizing organizational accounts of function. *Applied Ontology, 6*(2), 105–124.

Atkins, P. W. (1984). *The second law*. New York: Freeman.

Barandiaran, X. (2008). *Mental life. A naturalized approach to the autonomy of cognitive agents*. Ph.D. dissertation, University of the Basque Country (UPV/EHU).

Barandiaran, X., & Egbert, M. D. (2013). Norm-establishing and norm-following in autonomous agency. *Artificial Life Journal, 20*(1), 5–28.

A. Moreno, M. Mossio, *Biological Autonomy*, History, Philosophy and Theory of the Life Sciences 12, DOI 10.1007/978-94-017-9837-2

Barandiaran, X., & Moreno, A. (2006). On what makes certain dynamical systems cognitive. *Adaptive Behavior, 14*, 171–185.

Barandiaran, X., & Moreno, A. (2008). Adaptivity: from metabolism to behavior. *Adaptive Behavior, 16*(5), 325–344.

Barandiaran, X., Di Paolo, E., & Rohde, M. (2009). Defining agency. Individuality, normativity, asymmetry and spatio-temporality in action. *Journal of Adaptive Behavior, 17*(5), 367–386.

Barthal, I. B., Decety, J., & Mason, P. (2011). Empathy and pro-social behavior in rats. *Science, 334*(6061), 1427–1430.

Bechara, A. (2004). The role of emotion in decision-making: evidence from neurological patients with orbitofrontal damage. *Brain and Cognition, 55*, 30–40.

Bechtel, W. (2007). Biological mechanisms: organized to maintain autonomy. In F. Boogerd, F. Bruggeman, J. H. Hofmeyr, & H. V. Westerhoff (Eds.), *Systems biology. Philosophical foundations* (pp. 269–302). Amsterdam: Elsevier.

Bedau, M. A. (1992). Goal-directed systems and the good. *The Monist, 75*, 34–49.

Bedau, M. A. (1996). The nature of life. In M. Boden (Ed.), *The philosophy of artificial life* (pp. 332–357). New York: Oxford University Press.

Bell, G., & Mooers, A. O. (1997). Size and complexity among multicellular organisms. *Biological Journal of the Linnean Society, 60*, 345–363.

Benner, S. A. (1999). How small can a microorganism be? In *Size limits of very small microorganisms (Proceedings on a workshop)* (pp. 126–135). Washington, DC: National Academy Press.

Ben-Tabou de-Leon, S., & Davidson, E. H. (2007). Gene regulation: gene control network in development. *Annual Review of Biophysics and Biomolecular Structure, 36*, 191–212.

Berger, S. L., Kouzarides, T., Shiekhattar, R., & Shilatifard, A. (2009). An operational definition of epigenetics. *Genes and Development, 23*(7), 781–783.

Berleman, J., & Kirby, J. (2009). Deciphering the hunting strategy of a bacterial wolfpack. *FEMS Microbiology Reviews, 33*(5), 942–957.

Bernard, C. (1865). *Introduction á l'étude de la médecine expérimentale*. Paris: Baillière.

Bernard, C. (1878). *Leçons sur les phénomènes de la vie communs aux animaux et aux végétaux*. Paris: Baillière.

Bich, L. (2012). Complex emergence and the living organization: an epistemological framework for biology. *Synthese, 185*, 215–232.

Bich, L., Moreno, M., Mossio, M., & Ruiz-Mirazo, K. (forthcoming). Biological regulation: controlling the system from within.

Bickhard, M. H. (2000). Autonomy, function, and representation. *Communication and Cognition Artificial Intelligence, 17*(3–4), 111–131.

Bickhard, M. H. (2004). Process and emergence: normative function and representation. *Axiomathes – An International Journal in Ontology and Cognitive Systems, 14*, 121–155.

Bigelow, J., & Pargetter, R. (1987). Functions. *Journal of Philosophy, 84*, 181–196. Reprinted in Buller, D. J. (1999). *Function, selection, and design* (pp. 97–114). Albany: SUNY Press.

Bitbol, M. (2007). Ontology, matter and emergence. *Phenomenology and the Cognitive Science, 6*, 293–307.

Block, N. (2003). Do causal powers drain away? *Philosophy and Phenomenological Research, 67*(1), 133–150.

Bonner, J. T. (1988). *The evolution of complexity by means of natural selection*. Princeton: Princeton University Press.

Bonner, J. T. (1999). The origins of multicellularity. *Integrative Biology, 1*, 27–36.

Bonner, J. T. (2000). *First signals. The evolution of multicellular development*. Princeton: Princeton University Press.

Boogerd, F., Bruggeman, F., Hofmeyr, J.-H., & Westerhoff, H. V. (Eds.). (2007). *Systems biology. Philosophical foundations*. Amsterdam: Elsevier.

Boorse, C. (1976). Wright on functions. *Philosophical Review, 85*, 70–86.

Boorse, C. (1977). Health as a theoretical concept. *Philosophy of Science, 44*(4), 542–573.

Boorse, C. (1997). A rebuttal on health. In J. M. Humber & R. F. Almeder (Eds.), *What is Disease?* (pp. 1–134). Totowa, NJ: Humana Press.

Boorse, C. (2002). A rebuttal on functions. In A. Ariew, R. Cummins, & M. Perlman (Eds.), *Functions* (pp. 63–112). Oxford: Oxford University Press.

Bourgine, P., & Stewart, J. (2004). Autopoiesis and cognition. *Artificial Life, 10*, 327–345.

Bourgine, P., & Varela, F. J. (1992). *Toward a practice of autonomous systems*. Cambridge, MA: MIT Press/Bradford Books.

Broad, C. D. (1925). *The mind and its place in nature*. London: Routledge and Kegan Paul.

Brown, T. S. (1911). The intrinsic factors in the act of progression in the mammal. *Proceeding of the Royal Society London, Series B, 84*, 308–319.

Budin, I., & Szostak, J. W. (2011). Physical effects underlying the transition from primitive to modern cell membranes. *PNAS, 108*(13), 5249–5254.

Buller, D. J. (1999). *Function, selection, and design*. Albany: SUNY Press.

Bullock, T. H., Bennett, M. V. L., Johnston, D., Josephson, R., Marder, E., & Fields, R. D. (2005). The neuron doctrine, redux. *Science, 4–310*(5749), 791–793.

Burge, T. (2009). Primitive agency and natural norms. *Philosophy and Phenomenological Research, 79*(2), 251–278.

Burge, T. (2010). *Origins of objectivity*. Oxford: Oxford University Press.

Buss, L. W. (1987). *The evolution of individuality*. Princeton: Princeton University Press.

Cahiers du CREA. (1985). *Histoires de cybernetique* (Vol. 7). Paris: Ecole Polytechnique.

Campbell, D. T. (1974). Downward causation in hierarchically organized biological systems. In F. J. Ayala & T. Dobzhansky (Eds.), *Studies in the philosophy of biology* (pp. 179–186). Berkeley/Los Angeles: University of California Press.

Campbell, R. J., & Bickhard, M. H. (2011). Physicalism, emergence and downward causation. *Axiomathes, 21*(1), 33–56. Quotations from the online version: http://www.lehigh.edu/~mhb0/physicalemergence.pdf.

Campbell, E. L., Summers, M. L., Christman, H., Martin, M. E., & Meeks, J. C. (2007). Global gene expression patterns of *Nostoc punctiforme* in steady-state dinitrogen-grown heterocyst-containing cultures and at single time points during the differentiation of akinetes and hormogonia. *Journal of Bacteriology, 189*, 5247–5256.

Canfield, J. (1964). Teleological explanation in biology. *British Journal for the Philosophy of Science, 14*, 285–295.

Cannon, W. B. (1929). Organisation for physiological homeostasis. *Physiological Reviews, 9*(3), 399–431.

Cárdenas, M. L., Letelier, J. C., Gutierrez, C., Cornish-Bowden, A., & Soto-Andrade, J. (2010). Closure to efficient causation, computability and artificial life. *Journal of Theoretical Biology, 263*(1), 79–92.

Carroll, S. B. (2001). Chance and necessity: the evolution of morphological complexity and diversity. *Nature, 409*, 1102–1109.

Centler, F., & Dittrich, P. (2007). Chemical organizations in atmospheric photochemistries: a new method to analyze chemical reaction networks. *Planetary and Space Science, 55*(4), 413–428.

Chandler, J. L. R., & Van De Vijver, G. (Eds.). (2000). *Closure: emergent organizations and their dynamics* (Vol. 901). New York: Annals of the New York Academy of Science.

Chandresekhar, S. (1961). *Hydrodynamic and hydromagnetic stability*. Oxford: Clarendon.

Chiel, H., & Beer, R. (1997). The brain has a body: adaptive behaviour emerges from interactions of nervous system, body and environment. *Trends Neuroscience, 20*, 553–557.

Christensen, W. D. (2007). Volition and cognitive control. In D. Spurrett, H. Kincaid, L. Stephens, & D. Ross (Eds.), *Distributed cognition and the will: individual volition and social context* (pp. 255–287). Cambridge, MA: MIT Press.

Christensen, W. D. (2012). Natural sources of normativity. *Studies in History and Philosophy of Science Part C, 43*(1), 104–112.

Christensen, W. D., & Bickhard, M. H. (2002). The process dynamics of normative function. *The Monist, 85*(1), 3–28.

Christensen, W. D., & Hooker, C. A. (2000). An interactivist-constructivist approach to intelligence: self-directed anticipative learning. *Philosophical Psychology, 13*, 5–45.

Christman, H. D., Campbell, E. L., & Meeks, J. C. (2011). Global transcription profiles of the nitrogen stress response resulting in heterocyst or hormogonium development in *Nostoc punctiforme*. *Journal of Bacteriology, 193*(24), 6874–6886.

Clarke, E. (2011). The problem of biological individuality. *Biological Theory, 5*(4), 312–325.

Clarke, E. (2013). The multiple realizability of biological individuals. *The Journal of Philosophy, 8*, 413–435.

Cleland, C., & Chyba, C. (2007). *Does life have a definition?* In W. T. Sullivan & J. A. Baross (Eds.), *Planets and life: The emerging science of astrobiology* (Ch. 5, pp. 119–131). Cambridge: Cambridge University Press. Reprinted in C. E. Cleland & M. A. Bedau (Eds.) *The nature of life: Classical and contemporary perspectives from philosophy and science*. Cambridge, 2010, pp. 326–339.

Collier, J. D. (2000). Autonomy and process closure as the basis for functionality. In J. L. R. Chandler & G. van der Vijver (Eds.), *Closure: emergent organisations and their dynamics* (pp. 280–290). New York: Annals of the New York Academy of Sciences.

Conrad, M. (1979). Bootstrapping on the adaptive landscape. *BioSystems, 11*, 167–182.

Costa, S., & Shaw, P. (2007). 'Open minded' cells: how cells can change fate. *Trends in Cell Biology, 17*, 101–106.

Craver, C. F. (2001). Role functions, mechanisms, and hierarchy. *Philosophy of Science, 68*, 53–74.

Craver, C. F., & Bechtel, W. (2007). Top-down causation without top-down causes. *Biology and Philosophy, 22*, 547–563.

Crick, F., & Koch, C. (1990). Towards a neurobiological theory of consciousness. *Seminars in the Neurosciences, 2*, 263–275.

Crutchfield, J. P. (1994). The calculi of emergence: computations, dynamics, and induction. *Physica D, 75*, 11–54.

Cummins, R. (1975). Functional analysis. *Journal of Philosophy, 72*, 741–765. Reprinted in Buller, D. J. (1999). *Function, selection, and design* (pp. 57–83). Albany: SUNY Press.

Cummins, R. (2002). Neo-teleology. In A. Ariew, R. Cummins, & M. Perlman (Eds.), *Functions* (pp. 157–172). Oxford: Oxford University Press.

Dagg, J. (2003). Ecosystem organization as side-effects of replicator and interactor activities. *Biology and Philosophy, 18*, 491–492. doi:10.1023/A:1024128115666.

Damasio, A. R. (1994). *Descartes' error. Emotion, reason and the human brain*. New York: G.P. Putnam's Sons.

Damasio, A. R. (1999). *The feeling of what happens*. New York: Harcourt Brace and Company.

Davidson, D. (1963). Actions, reasons, and causes. *The Journal of Philosophy, 60*(23), 685–700.

Davies, P. S. (1994). Troubles for direct proper functions. *Noûs, 28*, 363–381.

Davies, P. S. (2000). Malfunctions. *Biology and Philosophy, 15*, 19–38.

Davies, P. S. (2001). *Norms of nature. Naturalism and the nature of functions*. Cambridge, MA: MIT Press.

Dawkins, R. (1976). *The selfish gene*. New York: Oxford University Press.

Dawkins, R. (1982). *The extended phenotype: the long reach of the gene*. Oxford: Oxford University Press.

De Duve, C. (2007). Chemistry and selection. *Chemistry and Biodiversity, 4*(4), 574–583.

De Groot, R. S., Wilson, M. A., & Boumans, R. (2002). A typology for the classification, description and valuation of ecosystem functions, goods and services. *Ecological Economics, 41*, 393–408.

De Jaegher, H., & Di Paolo, E. (2007). Participatory sense-making: an enactive approach to social cognition. *Phenomenology and the Cognitive Sciences, 6*(4), 485–507.

De Jaegher, H., Di Paolo, E., & Gallagher, S. (2010). Can social interaction constitute social cognition? *Trends in Cognitive Science, 14*, 441–447.

Deamer, D. W. (1997). The first living systems: a bioenergetic perspective. *Microbiology and Molecular Biology Reviews, 61*(2), 239–261.

Deamer, D. W. (2008). Origins of life: how leaky were primitive cells? *Nature, 454*, 37–38.

Deamer, D. W. (2009). On the origin of systems. Systems biology, synthetic biology and the origin of life. *EMBO Reports, 10*(1), S1–S4.

Delancey, C. (2006). Ontology and teleofunctions: a defense and revision of the systematic account of teleological explanation. *Synthese, 150*, 69–98.

Dennett, D. (1986). *Kinds of minds: towards an understanding of consciousness*. New York: Basic Books.

Di Paolo, E. A. (2005). Autopoiesis, adaptivity, teleology, agency. *Phenomenology and the Cognitive Sciences, 4*(4), 429–452.

Di Paolo, E. A., & De Jaegher, H. (2012). The interactive brain hypothesis. *Frontiers in Human Neuroscience, 6*, 163.

Dobzhansky, T. (1973). Nothing in biology makes sense except in the light of evolution. *American Biology Teacher, 35*, 125–129.

Dretske, F. (1988). *Explaining behavior*. Cambridge, MA: MIT Press.

Duchesneau, F. (1974). Du modèle cartésien au modèle spinoziste de l'être vivant. *Canadian Journal of Philosophy, 3*(4), 539–562.

Edelman, G. (1987). *Neural Darwinism. The theory of neuronal group selection*. New York: Basic Books.

Edelman, G. M. (1989). *The remembered present: a biological theory of consciousness*. New York: Basic Books.

Edelman, G. M. (1992). *Brilliant air, brilliant fire: on the matter of mind*. New York: Basic Books.

Edelman, D. B., & Seth, A. K. (2009). Animal consciousness: a synthetic approach. *Trends in Neuroscience, 32*(9), 476–484.

Edelman, G. M., & Tononi, G. (2000). *Consciousness: how matter becomes imagination*. London: Penguin Books.

Edin, B. (2008). Assigning biological functions: making sense of causal chains. *Synthese, 161*, 203–218.

Eigen, M. (1971). Selforganization of matter and evolution of biological Macromolecules. *Naturwissenschaften, 58*(10), 465–523.

Emmeche, C. (1992). Life as an abstract phenomenon: is artificial life possible? In F. Varela & P. Bourgine (Eds.), *Toward a practice of autonomous systems* (pp. 466–474). Cambridge, MA: MIT Press.

Emmeche, C., Stjernfelt, F., & Køppe, S. (2000). Levels, emergence, and three versions of downward causation. In P. B. Andersen, C. Emmeche, N. O. Finnemann, & P. V. Christiansen (Eds.), *Downward causation. minds, bodies and matter* (pp. 13–34). Aarhus: Aarhus University Press.

Engel, A. K., & Singer, W. (2001). Temporal binding and the neural correlates of sensory awareness. *Trends in Cognitive Sciences, 5*(1), 16–25.

Ereshefsky, M., & Pedroso, M. (2013). Biological individuality: the case of biofilms. *Biology and Philosophy, 28*(2), 331–349.

Erickson, K., & Schulkin, J. (2003). Facial expressions of emotion: a cognitive neuroscience perspective. *Brain and Cognition, 52*(1), 52–60.

Eschenmosser, A. (2007). The search for the chemistry of life's origin. *Tetrahedron, 63*, 12821–12844.

Etxeberria, A., & Moreno, A. (2001). From complexity to simplicity: nature and symbols. *BioSystems, 60*(1–3), 149–157.

Etxeberria, A., & Umerez, J. (2006). Organización y organismo en la Biología Teórica ¿Vuelta al organicismo? *Ludus Vitalis, 26*, 3–38.

Farmer, J., Kauffman, S., & Packard, N. (1986). Autocatalytic replication of polymers. *Physica D, 22*, 50–67.

Fell, D. (1997). *Understanding the control of metabolism*. London: Portland University Press.

Flemming, H., & Wingender, J. (2010). The biofilm matrix. *Nature Reviews Microbiology, 8*, 623–633.

Folse, H. J., 3rd, & Roughgarden, J. (2010). What is an individual organism? A multilevel selection perspective. *The Quarterly Review of Biology, 85*(4), 447–472.

Fontana, W. (1992). Algorithmic chemistry. In C. G. Langton, C. Taylor, J. D. Farmer, & S. Rasmussen (Eds.), *Artificial life II* (pp. 159–209). Redwood City: Addison-Wesley.

Fontana, W., Wagner, G., & Buss, L. W. (1994). Beyond digital naturalism. *Artificial Life, 1*(1/2), 211–227.
Fox Keller, E. (2007). The disappearance of function from 'self-organizing systems'. In F. Boogerd, F. Bruggeman, J.-H. Hofmeyr, & H. V. Westerhoff (Eds.), *Systems biology: philosophical foundations* (pp. 303–317). Amsterdam: Elsevier.
Fox Keller, E. (2009). Self-organization, self-assembly, and the inherent activity of matter. In S. H. Otto (Ed.), *The Hans Rausing lecture 2009* (Uppsala University, Disciplinary Domain of Humanities and Social Sciences, Faculty of Arts, Department of History of Science and Ideas).
Fox Keller, E. (2010). It is possible to reduce biological explanations to explanations in chemistry and/or physics? In F. J. Ayala & R. Arp (Eds.), *Contemporary debates in philosophy of biology* (pp. 19–31). Chichester: Wiley.
Frankfurt, H. G. (1978). The problem of action. *American Philosophical Quarterly, 15*(2), 157–162.
Fraser, O., & Bugnyar, T. (2010). Do ravens show consolation? Responses to distressed others. *PLoS ONE, 5*(5), 1–8.
Fry, I. (2000). *The emergence of life on Earth: a historical and scientific overview*. London: Rutgers University Press.
Fuchs, T., & De Jaegher, H. (2009). Enactive intersubjectivity: participatory sense-making and mutual incorporation. *Phenomenology and the Cognitive Sciences, 8*, 465–486.
Ganti, T. (1975). Organization of chemical reactions into dividing and metabolizing units: the chemotons. *BioSystems, 7*, 15–21.
Ganti, T. (1973/2003). *The principles of life*. Oxford: Oxford University Press.
Gardner, A. (2009). Adaptation as organism design. *Biology Letters, 5*(6), 861–864.
Gardner, A., & Grafen, A. (2009). Capturing the superorganism: a formal theory of group adaptation. *Journal of Evolutionary Biology, 22*(4), 659–671.
Gaukroger, S., Schuster, J., & Sutton, J. (Eds.). (2000). *Descartes' natural philosophy*. London: Routledge.
Gayon, J. (2006). Les biologistes ont-ils besoin du concept de fonction? Perspective philosophique. *Comptes Rendus Palevol, 5*(3–4), 479–487.
Gayon, J., & de Ricqlès, A. (Eds.). (2010). *Les fonctions: des organismes aux artefacts*. Paris: Presses Universitaires de France.
Gerhart, J., & Kirschner, M. (1997). *Cells, embryos and evolution*. Malden: Blackwell Science.
Gil, R., Silva, F. J., Peretó, J., & Moya, A. (2004). Determination of the core of a minimal bacteria gene set. *Microbiology and Molecular Biology Reviews, 68*, 518–537.
Gilbert, S. F., & Sarkar, S. (2000). Embracing complexity: organicism for the 21st century. *Developmental Dynamics, 219*(1), 1–9.
Glansdorff, P., & Prigogine, I. (1971). *Thermodynamics of structure, stability and fluctuations*. London: Wiley.
Gloor, P. (1997). *The temporal lobe and limbic system*. New York: Oxford University Press.
Godfrey-Smith, P. (1994). A modern history theory of functions. *Noûs, 28*, 344–362. Reprinted in Buller, D. J. (1999). *Function, selection, and design* (pp. 199–220). Albany: SUNY Press.
Godfrey-Smith, P. (2009). *Darwinian populations and natural selection*. Oxford: Oxford University Press.
Godfrey-Smith, P. (2010). *Thinking like an octopus*. Harvard Gazette, Oct 21.
Godfrey-Smith, P. (2013): On being an octopus. Boston Review. https://www.bostonreview.net/books-ideas/peter-godfrey-smith-being-octopus
Godfrey-Smith, P. (2014). *Philosophy of biology*. Princeton: Princeton University Press.
Gommans, W., Mullen, S., & Maas, S. (2009). RNA editing: a driving force for adaptive evolution? *Bioessays, 31*(10), 1137–1145.
Gould, S. J. (1988). On replacing the idea of progress with an operational notion of directionality. In N. H. Nitecki (Ed.), *Evolutionary progress* (pp. 319–338). Chicago: University of Chicago Press.
Gould, S. J. (1994). The evolution of life on the Earth. *Scientific American,* Oct 85–91.9.

Gould, S. J. (2002). *The structure of evolutionary theory*. Cambridge, MA: Harvard University Press.

Gould, S. J., & Lewontin, R. C. (1979). The spandrels of San Marco and the Panglossian paradigm: a critique of the adaptationist programme. *Proceedings of the Royal Society of London B, 205*, 581–598.

Griesemer, J. R. (2002). What is “epi” about epigenetics? In L. Van Speybroeck, G. Van de Vijver, & D. De Waele (Eds.), *From epigenesis to epigenetics: the genome in context* (pp. 97–110). New York: New York Academy of Sciences.

Griesemer, J. R., & Szathmáry, E. (2009). Ganti’s chemoton model and life criteria. In S. Rasmussen, M. A. Bedau, L. Chen, D. Deamer, D. C. Krakauer, & N. H. Packard (Eds.), *Protocells: bridging nonliving and living matter* (pp. 481–512). Cambridge, MA: MIT Press.

Griffiths, P. E. (1993). Functional analysis and proper functions. *British Journal for the Philosophy of Science, 44*, 409–422. Reprinted in Buller, D. J. (1999). *Function, selection, and design* (pp. 143–158). Albany: SUNY Press.

Griffiths, P. E. (2001). Genetic information: a metaphor in search of a theory. *Philosophy of Science, 68*(3), 394–412.

Grosberg, R. K., & Strathmann, R. R. (2007). The evolution of multicellularity: a minor major transition. *Annual Review of Ecology, Evolution, Systematics, 38*, 621–654.

Hallmann, A. (2011). Evolution of reproductive development in the volvocine algae. *Sexual Plant Reproduction, 24*, 97–112.

Hanczyc, M., & Ikegami, T. (2010). Chemical basis for minimal cognition. *Artificial Life, 16*(3), 233–243.

Hardcastle, V. G. (2002). On the normativity of functions. In A. Ariew, R. Cummins, & M. Perlman (Eds.), *Functions* (pp. 144–156). Oxford: Oxford University Press.

Hartenstein, V. (2006). The neuroendocrine system in inveretbrates: a developmental and evolutionary perspective. *Endocrinology, 190*, 555–570.

Hempel, C. G. (1965). *Aspects of scientific explanation*. London: Collier Macmillan Publishers.

Hempel, C. G., & Oppenheim, P. (1948). Studies in the logic of explanation. *Philosophy of Science, 15*, 135–175.

Hertel, J., Lindemeyer, M., Missal, K., Fried, C., Tanzer, A., Flamm, C., Hofacker, I. L., & Stadler, P. F. (2006). The expansion of the metazoan microRNA repertoire. *BMC Genomics, 7*(1), 25.

Hickman, C., Roberts, L., & Larson, A. (2001). *Integrated principles of zoology*. Boston: MacGraw-Hill.

Hoffmeyer, J. (1996). *Signs of meaning in the universe*. Bloomington: Indiana University Press.

Hoffmeyer, J. (1998). Life: the invention of externalism. In Farre, G., & Oksala, T. (Eds.), *Emergency, complexity, hierarchy, organization*. Selected and Edited Papers from the ECHO III Conference. Acta Polytechnica Scandinavica 91, Espoo, pp. 187–196.

Hooker, C. (2004). Asymptotics, reduction and emergence. *British Journal for the Philosophy of Science, 55*(3), 435–479.

Hooker, C. (2009). Interaction and bio-cognitive order. *Synthese, 166*, 513–546.

Hooker, C. (2011). Introduction to philosophy of complex systems. Part A: Towards framing philosophy of complex systems. In C. A. Hooker (Ed.), *Philosophy of complex systems* (Handbook of the philosophy of science, Vol. 10, pp. 3–92). Amsterdam: North Holland/Elsevier.

Hooker, C. (2013). On the import of constraints in complex dynamical systems. *Foundations of Science, 18*, 757–780.

Hooker, C., & Christensen, W. (1999). The organisation of knowledge: beyond Campbell’s evolutionary epistemology. *Philosophy of Science, 66*(3), 237–249.

Horibe, N., Hanczyc, M., & Ikegami, T. (2011). Mode switching and collective behavior in chemical oil droplets. *Entropy, 13*(3), 709–719.

Hull, D. L. (1988). *Science as a process: an evolutionary account of the social and conceptual development of science*. Chicago: University of Chicago Press.

Huneman, P. (2006). Naturalizing purpose: from comparative anatomy to the “adventures of reason”. *Studies in History and Philosophy of Life Sciences, 37*(4), 621–656.

Huneman, P. (Ed.). (2007). *Understanding purpose? Kant and the philosophy of biology*. Rochester: University of Rochester Press.

Jacobs, D. K., Nakanishi, N., Yuan, D., Camara, A., Nichols, S. A., & Hartenstein, V. (2007). Evolution of sensory structures in basal metazoa. *Integrative and Comparative Biology, 47*(5), 712–723.

Jonas, H. (1966/2001). *The phenomenon of life: toward a philosophical biology*. Evanston: Northwestern University Press.

Juarrero, A. (1999). *Dynamics in action: intentional behavior as a complex system*. Cambridge, MA: MIT Press.

Juarrero, A. (2009). Top-down causation and autonomy in complex systems. In N. Murphy, G. Ellis, & T. O'Connor (Eds.), *Downward causation and the neurobiology of free will* (pp. 83–102). Berlin/Heidelberg: Springer.

Kacian, D. L., Mills, D. R., Kramer, F. R., & Spiegelman, S. (1972). A replicating RNA molecule suitable for a detailed analysis of extracellular evolution and replication. *Proceedings of the National Academy of Sciences of the United States of America, 69*(10), 3038–3042.

Kaiser, D. (2001). Building a multicellular organism. *Annual Review of Genetics, 35*, 103–123.

Kandel, E. R. (1995). *Essentials of neural science and behavior*. Norwalk: Appleton and Lange.

Kant, E. (1790/1987). *Critique of judgment*. Indianapolis: Hackett Publishing.

Kauffman, S. (1986). Autocatalytic sets of proteins. *Journal of Theoretical Biology, 119*, 1–24.

Kauffman, S. (2000). *Investigations*. Oxford: Oxford University Press.

Keijzer, F., van Duijn, M., & Lyon, P. (2013). What nervous systems do: early evolution, input–output, and the skin brain thesis. *Adaptive Behavior, 21*(2), 67–85.

Khushf, G. (2007). An agenda for future debate on concepts of health and disease. *Medicine, Health Care and Philosophy, 10*, 19–27.

Kim, J. (1993). *Supervenience and mind: selected philosophical essays*. Cambridge: Cambridge University Press.

Kim, J. (1997). Explanation, prediction and reduction in emergentism. *Intellectica, 25*(2), 45–57.

Kim, J. (1998). *Mind in a physical world*. Cambridge, MA: MIT Press.

Kim, J. (2003). Blocking causal drainage and other maintenance chores with mental causation. *Philosophy and Phenomological Research, 67*(1), 151–176.

Kim, J. (2006). Emergence: core ideas and issues. *Synthese, 151*(3), 547–559.

Kim, J. (2010). *Essays in the metaphysics of mind*. Oxford: Oxford University Press.

Kimura, M. (1968). Explanatory rate at the molecular level. *Nature, 217*, 624–626.

Kirk, D. L. (1998). *Volvox: molecular-genetic origins of multicellularity and cellular differentiation* (Developmental and cell biology series). Cambridge: Cambridge University Press.

Kirk, D. L. (2005). A twelve-step program for evolving multicellularity and a division of labor. *BioEssays, 27*, 299–310.

Kirk, M. M., Ransick, A., McRae, S. E., & Kirk, D. L. (1993). The relationship between cell size and cell fate in *Volvox carteri*. *Journal of Cell Biology, 123*, 191–208.

Kirschner, M., & Gerhart, J. (1998). Evolvability. *PNAS, 95*(15), 8420–8427.

Kitano, H. (2002). Computational systems biology. *Nature, 420*, 206–210.

Kitcher, P. (1993). Function and design. *Midwest Studies in Philosophy, 18*, 379–397.

Koufopanou, V. (1994). The evolution of soma in the Vovocales. *American Naturalist, 143*, 907–931.

Kozmik, Z., Ruzickova, J., Jonasova, K., Matsumoto, Y., Vopalensky, P., Kozmikova, I., Strnad, H., Kawamura, S., Piatigorsky, J., & Paces, V. (2008). Assembly of the cnidarian camera-type eye from vertebrate-like components. *Proceedings of the National Academy of Sciences of the United States of America, 105*, 8989–8993.

Krohs, U. (2010). Dys-, Mal- et Non-: L'autre face de la fontionnalité. In J. Gayon & A. De Ricqlès (Eds.), *Les fonctions: des organismes aux artefacts* (pp. 337–352). Paris: Presses Universitaires de France.

Krohs, U. (2011). Functions and fixed types: biological and other functions in the post-adaptationist era. *Applied Ontology, 6*(2), 125–139.

Kumar, K., Mella Herrera, A. R., & Golden, W. J. (2010). Cyanobacterial Heterocysts. Cold Spring Harb, *Perspect Biol*. doi: 10.1101/cshperspect.a000315.

Kupiec, J. J., & Sonigo, P. (2000). *Ni Dieu ni gene. Pour une autre théorie de l'hérédité*. Paris: Seuil.

Lagzi, I. (2013). Chemical robotics- chemotactic drug carriers. *Central European Journal of Medicine, 8*(4), 377–382.

Laland, K., Sterelny, K., Odling-Smee, J., Hopitt, W., & Uller, T. (2011). Cause and effect in biology revisited: Is Mayr's proximate-ultimate dichotomy still useful? *Science, 334*(6062), 1512–1516.

Laubichler, D., & Maienschein, J. (Eds.). (2007). *From embriology to evo-devo. A history of developmental evolution*. Cambridge, MA: MIT Press.

Laughlin, R., & Pines, D. (2000). The theory of everything. *Proceedings of the National Academy of Science of the United States of America, 97*, 28–31.

Laughlin, R., Pines, D., Schmalien, J., Stojkovic, B., & Wolynes, P. (2000). The middle way. *Proceedings of the National Academy of Science of the United States of America, 97*, 32–37.

Ledoux, J. (1996). *The emotional brain: the mysterious underpinnings of emotional life*. New York: Simon & Schuster.

Lengeler, J. W. (2000). Metabolic networks: a signal-oriented approach in cellular models. *Biological Chemistry, 381*, 911–920.

Letelier, J.-C., Marin, J., & Mpodozis, J. (2003). Autopoietic and (M, R)-systems. *Journal of Theoretical Biology, 222*, 261–272.

Letelier, J.-C., Soto-Andrade, J., Guiñez Abarzua, F., Cornish-Bowden, A., & Cardenas, M. L. (2006). Organizational invariance and metabolic closure: analysis in terms of (M, R)-systems. *Journal of Theoretical Biology, 238*, 949–961.

Levi-Montalcini, R. L. (1999). *La galassia mente*. Milano: Baldini and Castoldi.

Levy, A. (2011). Information in biology: a fictionalist account. *Noûs, 45*(4), 640–657.

Lewis, M. D. (2005). Bridging emotion theory and neurobiology through dynamic systems modeling. *Behavioral and Brain Sciences, 28*(2), 169–194.

Lewontin, R. C. (1970). The units of selection. *Annual Review of Ecology and Systematics, 1*, 1–18.

Llinás, R., Ribary, U., Contreras, D., & Pedroarena, C. (1998). The neuronal basis for consciousness. *Philosophical Transactions of the Royal Society of London, Series B, 353*, 1841–1849.

Lloyd Morgan, C. (1923). *Emergent evolution*. London: Williams and Norgate.

Longo, G., & Montevil, M. (2014). *Perspectives on organism: biological time, symmetries and singularities*. Heidelberg: Springer.

Losik, R., & Kaiser, D. (1997). Why and how bacteria communicate. *Scientific American, 276*(2), 68–73.

Luisi, P. L. (2006). *The emergence of life: from chemical origins to synthetic biology*. Cambridge: Cambridge University Press.

Luisi, P. L., Ferri, F., & Stano, P. (2006). Approaches to semi-synthetic minimal cells: a review. *Naturwissenschaften, 93*(1), 1–13.

Lyon, P. (2006). The biogenic approach to cognition. *Cognitive Processing, 7*(1), 11–29.

Manrubia, S., & Briones, C. (2007). Modular evolution and increase of functional complexity in RNA molecules. *RNA, 13*(1), 97–107.

Mansy, S., Schrum, J. P., Krishnamurthy, M., Tobé, S., Treco, D. A., & Szostak, J. W. (2008). Template directed synthesis of a genetic polymer in a model protocell. *Nature, 454*, 122–126.

Martin, W., & Russell, M. J. (2003). On the origins of cells: a hypothesis for the evolutionary transitions from abiotic geochemistry to chemoautotrophic prokaryotes, and from prokaryotes to nucleated cells. *Philosophical Transaction: Biological Science, 358*, 59–85.

Mattick, J. (2004). The hidden genetic program of complex organisms. *Scientific American, 291*(4), 60–67.

Maturana, H., & Varela, F. (1973). *De máquinas y seres Vivos – Una teoría sobre la organización biológica*. Santiago de Chile: Editorial Universitaria S.A.

Maturana, H., & Varela, F. (1980). *Autopoiesis and cognition. The realization of the living*. Dordrecht: Reidel Publishing.

Maturana, H., & Varela, F. (1987). *The tree of knowledge: the biological roots of human understanding*. Boston: Shambhala Publications.

Maynard Smith, J. (1986). *The problems of biology*. Oxford: Oxford University Press.

Maynard Smith, J., & Szathmary, E. (1995). *Major transitions in evolution*. Oxford: Oxford University Press.

Maynard Smith, J., Burian, R., Kauffman, S., Alberch, P., Campbell, J., Goodwin, B., Lande, R., Raup, D., & Wolpert, L. (1985). Developmental constraints and evolution. *Quarterly Review of Biology, 60*, 265–287.

Mayr, E. (1961). Cause and effect in biology: kinds of causes, predictability, and teleology are viewed by a practicing biologist. *Science, 134*, 1501–1506.

Mayr, E. (2004). *What makes biology unique? Considerations on the autonomy of a scientific discipline*. Cambridge: Cambridge University Press.

McAdams, H. H., & Arkin, A. (1997). Stochastic mechanisms in gene expression. *PNAS, 94*(3), 814–819.

McFall-Ngai, M. J. (1999). Consequences of evolving with bacterial symbionts: insights from the Squid-Vibrio Associations. *Annual Review of Ecology and Systematics, 30*, 235–256.

McGann, M., & De Jaegher, H. (2009). Self-other contingencies: enacting social perception. *Phenomenology and the Cognitive Sciences, 8*, 417–437.

McLaughlin, B. P. (1992). The rise and fall of British emergentism. In A. Beckermann, H. Flohr, & J. Kim (Eds.), *Emergence or reduction? Essays on the prospects of nonreductive physicalism* (pp. 49–93). Berlin: Walter de Gruyter.

McLaughlin, P. (2001). *What functions explain. Functional explanation and self-reproducing systems*. Cambridge: Cambridge University Press.

McLaughlin, P. (2009). Functions and norms. In U. Krohs & P. Kroes (Eds.), *Functions in biological and artificial worlds. Comparative philosophical perspectives* (pp. 93–102). Cambridge, MA: MIT Press.

McMullin, B. (1997). SCL: an artificial chemistry in Swarm. *Santa Fe Institute Working Paper* (SFI).

McMullin, B., & Varela, F. J. (1997). Rediscovering computational autopoiesis. In P. Husbands & I. Harvey (Eds.), *Proceedings of the fourth European conference on artificial life* (pp. 38–47). Cambridge, MA: MIT Press.

Melander, P. (1997). *Analyzing functions. An essay on a fundamental notion in biology*. Stockholm: Almkvist & Wiksell International.

Meyer, L. M. N., Bomfim, G. C., & El-Hani, C. N. (2013). How to understand the gene in the 21st century. *Science Education, 22*(2), 345–374.

Michod, R. E. (1999). *Darwinian dynamics: evolutionary transitions in fitness and individuality*. Princeton: Princeton University Press.

Michod, R. E. (2005). On the transfer of fitness from the cell to the multicellular organism. *Biology and Philosophy, 20*, 967–987.

Mill, J. S. (1843). *A system of logic*. London: Parker.

Miller, R. V. (1998). Bacterial gene swapping in nature. *Scientific American, 278*(1), 66–71.

Miller, M. B., & Bassler, B. L. (2001). Quorum sensing in bacteria. *Annual Review of Microbiology, 55*, 165–199.

Millikan, R. G. (1984). *Language, thought, and other biological categories*. Cambridge, MA: MIT Press.

Millikan, R. G. (1989). In defense of proper functions. *Philosophy of Science, 56*, 288–302.

Millikan, R. G. (1993). Propensities, exaptations, and the brain. In R. G. Millikan (Ed.), *White queen psychology and other essays for Alice* (pp. 31–50). Cambridge, MA: MIT Press.

Millikan, R. G. (2002). Biofunctions: two paradigms. In A. Ariew, R. Cummins, & M. Perlman (Eds.), *Functions* (pp. 113–143). Oxford: Oxford University Press.

Mills, D. R., Peterson, R. L., & Spiegelman, S. (1967). An extracellular Darwinian experiment with a self-duplicating nucleic acid molecule. *Proceedings of the National Academy of Sciences of the United States of America, 58*(1), 217–224.

Monod, J. (1970). *Le hasard et la nécessité. Essai sur la philosophie naturelle de la biologie moderne*. Paris: éditions du Seuil.

Montévil, M., & Mossio, M. (2015). Biological organisation as closure of constraints. *Journal of Theoretical Biology, 372*, 179–191.

Morán, F., Moreno, A., Montero, F., & Minch, E. (1997). Further steps towards a realistic description of the essence of life. In C. G. Langton & K. Shimohara (Eds.), *Artificial life V (Proceedings of the 5th international workshop on the synthesis and simulation of living systems)* (pp. 255–263). London: MIT Press.

Morange, M. (2003). La vie expliquée ? 50 ans après la double hélice. Paris: Odile Jacob. Eng trans. (2008). *Life explained*. New Haven: Yale University Press.

Moreno, A. (2007). A systemic approach to the origin of biological organization. In F. Boogerd, J. H. Bruggeman, H. V. Hofmeyr, & H. Westerhoff (Eds.), *Systems biology. Philosophical foundations* (pp. 243–268). Dordrecht: Elsevier.

Moreno, A., & Etxeberria, A. (2005). Agency in natural and artificial systems. *Artificial Life, 11*(1–2), 161–175.

Moreno, A., & Fernández, J. (1990). Structural limits for evolutive capacities in molecular complex systems. *Biology Forum, 83*(2/3), 335–347.

Moreno, A., & Lasa, A. (2003). From basic adaptivity to early mind. *Evolution and Cognition, 9*(1), 12–30.

Moreno, A., & Ruiz Mirazo, K. (1999). Metabolism and the problem of its universalization. *BioSystems, 49*(1), 45–61.

Moreno, A., & Ruiz-Mirazo, K. (2009). The problem of the emergence of functional diversity in prebiotic evolution. *Biology and Philosophy, 24*(5), 585–605.

Moreno, A., & Umerez, J. (2000). Downward causation at the core of living organization. In P. B. Anderson, C. Emmeche, N. O. Finnemann, & P. V. Christiansen (Eds.), *Downward causation* (pp. 99–117). Aarhus: Aarhus University Press.

Moreno, A., Etxeberria, A., & Umerez, J. (1994). Universality without matter? In R. A. Brooks & P. Maes (Eds.), *Artificial life IV (Proceedings of the 4th international workshop on the synthesis and simulation of living systems)* (pp. 406–410). London: MIT Press.

Moreno, A., Etxeberria, A., & Umerez, J. (2008). The autonomy of biological individuals and artificial models. *BioSystems, 91*(2), 309–319.

Moreno, A., Ruiz-Mirazo, K., & Barandiaran, X. E. (2011). The impact of the paradigm of complexity on the foundational frameworks of biology and cognitive science. In C. A. Hooker, D. V. Gabbay, P. Thagard, & J. Woods (Eds.), *Handbook of the philosophy of science* (Philosophy of complex systems, pp. 311–333). Amsterdam: Elsevier.

Moreno, A., Umerez, J., & Ibáñez, J. (1997). Cognition and life. *The Autonomy of Cognition Brain & Cognition, 34*(1), 107–129. Academic Press.

Morowitz, H. J. (1968). *Energy flow in biology*. New York: Academic Press.

Morowitz, H. J. (1992). *Beginnings of cellular life*. New Haven: Yale University Press.

Mossio, M. (2013). Closure, causal. In W. Dubitzky, O. Wolkenhauer, K.-H. Cho, & H. Yokota (Eds.), *Encyclopedia of systems biology* (pp. 415–418). New York: Springer.

Mossio, M., & Moreno, A. (2010). Organizational closure in biological organisms. *History and Philosophy of the Life Sciences, 32*(2–3), 269–288.

Mossio, M., Bich, L., & Moreno, A. (2013). Emergence, closure and inter-level causation in biological systems. *Erkenntnis, 78*(2), 153–178.

Mossio, M., Longo, G., & Stewart, J. (2009a). A computable expression of closure to efficient causation. *Journal of Theoretical Biology, 257*(3), 489–498.

Mossio, M., Saborido, C., & Moreno, A. (2009b). An organizational account of biological functions. *British Journal for the Philosophy of Science, 60*, 813–841.

Nagel, E. (1961). *The structure of science*. London: Routledge & Kegan Paul.

Nagel, E. (1977). Teleology revisited. *Journal of Philosophy, 74*, 261–301.

Neander, K. (1980). Teleology in biology. *Paper presented to the AAP conference*, Christchurch, New Zealand.

Neander, K. (1991). Function as selected effects: the conceptual analyst's defense. *Philosophy of Science, 58*, 168–184.
Neander, K. (1995). Misrepresenting and malfunctioning. *Philosophical Studies, 79*, 109–141.
Nedelcu, A. M., & Michod, R. E. (2004). Evolvability, modularity, and individuality during the transition to multicellularity in volvocalean green algae. In G. Wagner & G. Schlosser (Eds.), *Modularity in development and evolution* (pp. 468–489). Chicago: University of Chicago Press.
Neihardt, F. (Ed.). (1996). *Escherichia coli and salmonella: cellular and molecular biology*. Washington, DC: American Society for Microbiology.
Newell, A. (1980). Physical symbol systems. *Cognitive Science, 4*, 135–183.
Nicolis, G., & Prigogine, I. (1977). *Self-organisation in non-equilibrium systems: from dissipative structures to order through fluctuation*. New York: Wiley.
Nilsson, S., & Holmgren, S. (Eds.). (1994). *Comparative physiology and evolution of the autonomic nervous system*. Chur: Harwood Academic Publishers.
Nissen, L. (1980). Nagel's self-regulation analysis of teleology. *Philosophical Forum, 12*, 128–138.
Nunes, N., Moreno, A., & El Hani, C. (2014). Function in ecology: an organizational approach. *Biology and Philosophy, 29*(1), 123–141.
Oehlenschläger, F., & Eigen, M. (1997). 30 years later—a new approach to Sol Spiegelman's and Leslie Orgel's in vitro evolutionary studies. Dedicated to Leslie Orgel on the occasion of his 70th birthday. *Origins of Life and Evolution of the Biosphere, 27*(5–6), 437–457.
Oliveri, P., Tu, Q., & Davidson, E. H. (2008). Global regulatory logic for specification of an embryonic cell lineage. *Proceedings of the National Academy of Sciences of the United States of America, 105*, 5955–5962.
Oyama, S. (2002). The nurturing of natures. In A. Grunwald, M. Gutmann, & E. Neumann-Held (Eds.), *On human nature. Anthropological, biological and philosophical foundations* (pp. 163–170). New York: Springer.
Panksepp, J. B., & Lahvis, G. P. (2011). Rodent empathy and affective neuroscience. *Neuroscience and Biobehavioral Reviews, 35*, 1864–1875.
Pattee, H. H. (1972). Laws and constraints, symbols and languages. In C. H. Waddington (Ed.), *Towards a theoretical biology* (Vol. 4). Edinburgh: Edinburgh University Press.
Pattee, H. H. (1973). The physical basis and origin of hierarchical control. In H. H. Pattee (Ed.), *Hierarchy theory* (pp. 73–108). New York: Braziller.
Pattee, H. H. (1977). Dynamic and linguistic modes of complex systems. *International Journal of General Systems, 3*, 259–266.
Pattee, H. H. (1982). Cell psychology: an evolutionary approach to the symbol-matter problem. *Cognition and Brain Theory, 4*, 325–341.
Pattee, H. H. (2007). Laws, constraints, and the modeling relation – History and interpretations. *Chemistry and Biodiversity, 4*, 2272–2295.
Pepper, J. W., & Herron, M. D. (2008). Does biology need an organism concept? *Biological Reviews, 83*(4), 621–627.
Pereto, J. (2005). Controversies on the origin of life. *International Microbiology, 8*, 23–31.
Perlman, R. L. (2000). The concept of the organism in physiology. *Theory in Bioscience, 119*, 174–186.
Pessoa, L. (2008). On the relationship between emotion and cognition nature reviews. *Nature Reviews Neuroscience, 9*(2), 148–158.
Peter, I., & Davidson, E. H. (2009). Genomic control of patterning. *International Journal of Developmental Biology, 53*, 707–716.
Peter, I., & Davidson, E. H. (2010). The endoderm gene regulatory network in sea urchin embryos up to mid-blastula stage. *Developmental Biology, 340*(2), 188–199.
Peter, I., & Davidson, E. H. (2011). A gene regulatory network controlling the embryonic specification of endoderm. *Nature, 474*, 635–639.
Piaget, J. (1967). *Biologie et connaissance*. Paris: Éditions de la Pléiade.

Piedrafita, G., Montero, F., Morán, F., Cárdenas, M. L., & Cornish-Bowden, A. (2010). A simple self-maintaining metabolic system: robustness, autocatalysis, bistability. *PLoS Computational Biology, 6*(8), e1000872.

Pigliucci, M., & Muller, G. (Eds.). (2010). *Evolution: the extended synthesis*. Cambridge, MA: MIT Press.

Porges, S. (1997). Emotion: an evolutionary by-product of the neural regulation of the autonomic nervous system. In C. S. Carter, B. Kirkpatrick, & I. I. Lederhendler (Eds.), *The integrative neurobiology of affiliation* (Vol. 807, pp. 62–67). New York: Annals of the New York Academy of Sciences.

Powers, W. T. (1973). *Behavior: the control of perception*. Chicago: Aldine de Gruyter.

Price, T., Qvarnström, A., & Irwin, D. (2003). The role of phenotypic plasticity in driving genetic evolution. *Proceedings of the Royal Society of London B, 270*, 1433–1440.

Prigogine, I. (1962). *Introduction to nonequilibrium thermodynamics*. New York: Wiley-Interscience.

Pross, A. (2003). The driving force for life's emergence: kinetic and thermodynamic considerations. *Journal of Theoretical Biology, 220*(3), 393–406.

Queller, D. C., & Strassmann, J. E. (2009). Beyond society: the evolution of organismality. *Philosophical Transactions Royal Society, B: Biological Sciences, 364*, 3143–3155.

Raff, R. (1996). *The shape of life*. Chicago: University of Chicago Press.

Rasmussen, S., Bedau, M. A., Liaohai, C., Deamer, D., Krakauer, D. C., Packhard, N. H., & Stadler, P. F. (Eds.). (2008). *Protocells: bridging nonliving and living matter*. Cambridge, MA: MIT Press.

Richards, R. J. (2000). Kant and Blumenbach on the Bildungstrieb: a historical misunderstanding. *Studies in History and Philosophy of Biology and Biomedical Science, 31*(1), 11–32.

Rokas, A. (2008). The origins of multicellularity and the early history of the genetic toolkit for animal development. *Annual Review of Genetics, 42*, 235–251.

Rosen, R. (1971). Some realizations of (M, R)-systems and their interpretation. *Bulletin of Mathematical Biophysics, 33*, 303–319.

Rosen, R. (1972). Some relational cell models: the metabolism-repair systems. In R. Rosen (Ed.), *Foundations of mathematical biology* (Vol. II, pp. 217–253). New York: Academic Press.

Rosen, R. (1973). On the dynamical realizations of (M, R)-systems. *Bulletin of Mathematical Biophysics, 35*, 1–9.

Rosen, R. (1978). *Fundamentals of measurement and representation of natural systems*. New York: North Holland.

Rosen, R. (1991). *Life itself. A comprehensive enquiry into the nature, origin and fabrication of life*. New York: Columbia University Press.

Rosenblueth, A., Wiener, N., & Bigelow, J. (1943). Behavior, purpose and teleology. *Philosophy of Science, 10*, 18–24.

Rosslenbroich, B. (2014). *On the origin of autonomy. A new look at the major transitions in evolution*. Cham: Springer.

Ruiz-Mirazo, K. (2001). *Condiciones físicas para la aparición de sistemas autónomos con capacidades evolutivas abiertas*. Ph.D. dissertation, University of the Basque Country.

Ruiz-Mirazo, K. (2011). Protocell. In M. Gargaud, R. Amils, J. Cernicharo Quintanilla, H. J. Cleaves, W. M. Irvine, D. Pinti, & M. Viso (Eds.), *Encyclopedia of astrobiology* (Vol. 3, pp. 1353–1354). Heidelberg: Springer.

Ruiz Mirazo, K., Briones, C., & Escosura, A. (2013). Prebiotic systems chemistry: new perspectives for the origins of life. *Chemical Reviews, 114*(1), 285–366. 2014.

Ruiz-Mirazo, K., & Mavelli, F. (2007). Simulation model for functionalized vesicles: lipid-peptide integration in minimal protocells. In F. Almeida e Costa, L. M. Rocha, I. Harvey, & A. Coutinho (Eds.), *ECAL 2007*, Lisbon, Portugal, 10–14 Sept 2007. Proceedings (Lecture notes in computer science 4648, pp. 32–41). Heidelberg: Springer.

Ruiz-Mirazo, K., & Mavelli, F. (2008). Towards 'basic autonomy': stochastic simulations of minimal lipid- peptide cells. *BioSystems, 91*(2), 374–387.
Ruiz-Mirazo, K., & Moreno, A. (2004). Basic autonomy as a fundamental step in the synthesis of life. *Artificial Life, 10*(3), 235–259.
Ruiz-Mirazo, K., & Moreno, A. (2006). On the origins of information and its relevance for biological complexity. *Biological Theory, 1*(3), 227–229.
Ruiz-Mirazo, K., & Moreno, A. (2009). New century biology could do with a universal definition of life. In G. Terzis & R. Arp (Eds.), *Information and living systems: essays in philosophy of biology*. Cambridge, MA: MIT Press.
Ruiz-Mirazo, K., & Moreno, A. (2012). Autonomy in evolution: from minimal to complex life. *Synthese, 185*(1), 21–52.
Ruiz-Mirazo, K., Etxeberria, A., Moreno, A., & Ibañez, J. (2000). Organisms and their place in biology. *Theory in Biosciences, 119*, 43–67.
Ruiz-Mirazo, K., Peretó, J., & Moreno, A. (2004). A universal definition of life: autonomy and open-ended evolution. *Origins of Life and Evolution of the Biosphere, 34*(3), 323–346.
Ruiz-Mirazo, K., Umerez, J., & Moreno, A. (2008). Enabling conditions for open-ended evolution. *Biology and Philosophy, 23*(1), 67–85.
Ruse, M. (1971). Functional statements in biology. *Philosophy of Science, 38*, 87–95.
Saborido, C. (2012). *Funcionalidad y organización en biología. Reformulación del concepto de función biológica desde una perspectiva organizacional.* Ph.D. dissertation. University of the Basque Country.
Saborido, C., Mossio, M., & Moreno, A. (2011). Biological organization and cross-generation functions. *The British Journal for the Philosophy of Science, 62*, 583–606.
Saborido, C., & Moreno, A. (2015). Biological pathology from an organizational perspective. *Journal of Theoretical Medicine and Bioethics, 36*(1), 86–95.
Saborido, C., Moreno, A., González-Moreno, M., & Hernández, J. C. (2014). Organizational malfunctions and the notions of health and disease. In M. Lemoine & E. Giroux (Eds.), *Naturalism in philosophy of health: issues, limits and implications.* In Press, Springer.
Salmon, W. C. (1998). *Causality and explanation.* Oxford: Oxford University Press.
Santelices, B. (1999). How many kinds of individual are there? *Trends in Ecology & Evolution, 14*(4), 152–155.
Sartenaer, O. (2013). *Émergence et réduction. Analyse épistémologique de la dynamique des systèmes naturels.* Ph.D. dissertation, Université Catholique de Louvain.
Schlosser, G. (1998). Self-re-production and functionality: a systems-theoretical approach to teleological explanation. *Synthese, 116*, 303–354.
Schramme, T. (2007). A qualified defence of a naturalist theory of health. *Medicine, Health Care and Philosophy, 10*, 11–17.
Schrödinger, E. (1944). *What is life? The physical aspect of the living cell.* Cambridge: Cambridge University Press.
Science. (2002). *Special Issue on Systems Biology, 295*, 5560, 1589–1780.
Segré, D., & Lancet, D. (2000). Composing life. *EMBO Reports, 1*(3), 217–222.
Segré, D., Ben-Eli, D., Deamer, D., & Lancet, D. (2001). The lipid world. *Origins of Life and Evolution of the Biosphere, 31*, 119–145.
Seipel, K., & Schmid, V. (2005). Evolution of striated muscle: Jellyfish and the origin of triploblasty. *Developmental Biology, 282*, 14–26.
Sengupta, S., Ibele, M., & Sen, A. (2012). Fantastic voyage: designing self-powered nanorobots. *Angewandte Chemie International Edition, 51*(34), 8434–8445.
Seth, A. K. (2009). Brain mechanisms of consciousness in humans and other animals. In S. Atlason & Þ. Helgadóttir (Eds.), *Veit efnið af andanum* (pp. 47–72). Reykjavík: University of Iceland Press.
Shani, I. (2012). Setting the bar for cognitive agency: or how minimally autonomous can an autonomous agent be? *New Ideas in Psychology, 31*(2), 151–165.
Shapiro, J. A. (1998). Thinking about bacterial populations as multicellular organisms. *Annual Review of Microbiology, 52*, 81–104.

Shepherd, G. M. (1994). *Neurobiology* (3rd ed.). New York: Oxford University press.

Sherrington, C. S. (1947). *The integrative action of the nervous system* (2nd ed.). New Haven: Yale University Press.

Sherwood, L. (1997). *Human physiology*. Belmont: Wadsworth Publishing Company.

Silbersten, M., & McGeever, J. (1999). The search for ontological emergence. *Philosophical Quarterly, 50*(195), 182–200.

Simon, H. (1969). *The sciences of the artificial*. Cambridge, MA: MIT Press.

Simons, P. J. (1981). The role of electricity in plant movements. *New Phytologist, 87*, 11–37.

Skewes, J., & Hooker, C. (2009). Bio-agency and the problem of action. *Biology and Philosophy, 24*(3), 283–300.

Sloan, P. (2002). Preforming the categories: eighteenth–century generation theory and the biological roots of Kant's a–priori. *Journal of the History of Philosophy, 40*, 229–253.

Smithers, T. (1997). Autonomy in robots and other agents. *Brain and Cognition, 34*, 88–106.

Sober, E. (Ed.). (2006). *Conceptual issues in evolutionary biology*. Cambridge, MA: MIT Press.

Soto, A. M., Sonnenschein, C., & Miquel, P. A. (2008). On physicalism and downward causation in developmental and cancer biology. *Acta Biotheoretica, 56*, 257–274.

Sperry, R. W. (1969). A modified concept of consciousness. *Psychological Review, 76*(6), 532–536.

Spinoza, B. (1677/2002). *Ethics*. In Complete works translated by Samuel Shirley and others. Indianapolis: Hackett Publication Company.

Srere, P. A. (1984). Why are enzymes so big? *Trends in Biochemical Sciences, 9*, 387–390.

Stephan, A. (1992). Emergence – a systematic view on its historical facets. In A. Beckermann, H. Flohr, & J. Kim (Eds.), *Emergence or reduction? Essays on the prospects of nonreductive physicalism* (pp. 25–48). Berlin: Walter de Gruyter.

Sterelny, K., & Griffiths, P. (1999). *Sex and death. An introduction to philosophy of biology*. Chicago: The University of Chicago Press.

Stewart, J. (1996). Cognition=life: implications for higher-level cognition. *Behavioral Processes, 35*, 311–326.

Storer, T., Usinger, R., Stebbins, R., & Nybakken, J. (1979). *General zoology*. New York: McGraw-Hill.

Strassmann, J. E., & Queller, D. C. (2010). The social organism: congresses, parties and committees. *Evolution, 64*(3), 605–616.

Szathmary, E. (2000). The evolution of replicators. *Philosophical Transactions of the Royal Society of London B, 355*, 1669–1676.

Szathmary, E. (2006). The origin of replicators and reproducers. *Philosophical Transactions of the Royal Society B., 361*, 1761–1776.

Szathmary, E., & Maynard Smith, J. (1997). From replicators to reproducers: the first major transitions leading to life. *Journal of Theoretical Biology, 187*, 555–571.

Taft, R. J., Pheasant, M., & Mattick, J. S. (2007). The relationship between non-protein-coding DNA and eukaryotic complexity. *BioEssays, 29*, 288–297.

Teller, P. (1986). Relational holism and quantum mechanics. *The British Journal for the Philosophy of Science, 37*, 71–81.

Thompson, E. (2007). *Mind in life. Biology, phenomenology, and the sciences of mind*. Cambridge, MA: Harvard University Press.

Tononi, G., & Edelman, G. M. (1998). Consciousness and complexity. *Science, 262*, 1846–1851.

Tononi, G., Sporns, O., & Edelman, G. M. (1999). Measures of degeneracy and redundancy in biological networks. *Proceedings of the National Academy of Sciences USA, 96*, 3257–3262.

Trewavas, A. (2003). Aspects of plant intelligence. *Annals of Botany, 92*, 1–20.

Turroni, F., Ribbera, A., Foroni, E., van Sinderen, D., & Ventura, M. (2008). Human gut microbiota and bifidobacteria: from composition to functionality. *Antonie van Leeuwenhoek, 94*(1), 35–50.

Umerez, J. (1994). *Jerarquías Autónomas. Un estudio sobre el origen y la naturaleza de los procesos de control y de formación de niveles en sistemas naturales complejos*. Ph.D. dissertation. Donostia/San Sebastian: University of the Basque Country.

Umerez, J. (1995). Semantic closure: a guiding notion to ground artificial life. *Advances in artificial life, 929*, 77–94.

Umerez, J. (2001). Howard Pattee's theoretical biology—a radical epistemological stance to approach life, evolution and complexity. *Biosystems, 60*(1), 159–177.

Umerez, J., & Mossio, M. (2013). Constraint. In W. Dubitzky, O. Wolkenhauer, K.-H. Cho, & H. Yokota (Eds.), *Encyclopedia of systems biology* (pp. 490–493). New York: Springer.

Van Dujin, M., Keijzer, F., & Franjen, D. (2006). Principles of minimal cognition: casting cognition as sensorimotor coordination. *Adaptive Behavior, 14*(2), 157–170.

Van Gulick, R. (2001). Reduction, emergence and other recent options on the mind/body problem: a philosophical overview. *Journal of Consciousness Studies, 8*(9–10), 1–34.

Varela, F. J. (1979). *Principles of biological autonomy*. New York: North Holland.

Varela, F. J. (1981). Autonomy and autopoiesis. In G. Roth & H. Schwegler (Eds.), *Self-organizing systems: an interdisciplinary approach* (pp. 14–24). Frankfurt/New York: Campus Verlag.

Varela, F. J. (1997). Patterns of life: intertwining identity and cognition. *Brain and Cognition, 34*, 72–87.

Varela, F. J., Maturana, H., & Uribe, R. (1974). Autopoiesis: the organisation of living systems, its characterization and a model. *BioSystems, 5*, 187–196.

Vasas, V., Fernando, C., Santos, M., & Kauffman, S. (2012). Evolution before genes. *Biol Direct,* 7–1.

Vicente, A. (2011). Current physics and 'the physical'. *The British Journal for the Philosophy of Science, 62*(2), 393–416.

Vieira, F. S., & El-Hani, C. (2008). Downward determination: a philosophical step in the way to a dynamic account of emergence. *Cybernetics and Human Knowing, 15*(3–4), 145–147.

Virchow, R. (1858/1978). *Die Cellularpathologie in ihrer Begründung auf physiologische und pathologische Gewebelehre.* English translation, Cellular pathology. special ed., 204–207. John Churchill London, UK.

Vogel, S. (1988). *Life's devices: the physical world of animals and plants*. Princeton: Princeton University Press.

Von Uexkull, J. (1982/1940). The theory of meaning. *Semiotica, 42*(1),25–82.

Waddington, C. H. (1968–1972). *Towards a theoretical biology*. 4 vols. Edinburgh: Edinburgh University Press.

Wagner, G. P., & Altenberg, L. (1996). Complex adaptations and the evolution of evolvability. *Evolution, 50*(3), 967–976.

Walsh, D. M. (1996). Fitness and function. *British Journal for the Philosophy of Science, 47*, 553–574.

Walsh, D. M., & Ariew, A. (1996). A taxonomy of functions. *Canadian Journal of Philosophy, 26*, 493–514. Reprinted in Buller, D. J. (1999). *Function, selection, and design*(pp. 257–279). Albany: SUNY Press.

Ward, P., & Brownlee, D. (2004). *Rare earth: why complex life is uncommon in the universe*. New York: Copernicus Books.

Weber, A., & Varela, F. (2002). Life after Kant: natural purposes and the autopoietic foundations of biological individuality. *Phenomenology and the Cognitive Sciences, 1*, 97–125.

Wicken, J. S. (1987). *Evolution, thermodynamics and information. Extending the Darwinian program*. Oxford: Oxford University Press.

Wilson, J. (1999). *Biological individuality. The individuation and persistence of living entities.* Cambridge: Cambridge University Press.

Wilson, J. (2000). Ontological butchery: organism concepts and biological generalizations. *Philosophy of Science, 67*, 301–311.

Wilson, D. S., & Sober, E. (1989). Reviving the superorganism. *Journal of Theoretical Biology, 136*, 337–356.

Wimsatt, W. (2002). Functional organisation, functional inference, and functional analogy. In A. Ariew, R. Cummins, & M. Perlman (Eds.), *Functions* (pp. 174–221). Oxford: Oxford University Press.

Winther, R. (2006). Parts and theories in compositional. *Biology and Philosophy, 21*(4), 471–499.

Woese, C. (2002). On the evolution of cells. *Proceedings of the National academy of Science, 99*(13), 8742–8747.

Wolfe, C. (2010). Do organisms have an ontological status? *History and Philosophy of the Life Sciences, 32*(2–3), 195–232.
Wolpert, L., & Szathmary, E. (2002). Multicellularity: evolution and the egg. *Nature, 420*, 745.
Wouters, A. G. (2005). The function debate in philosophy. *Acta Biotheoretica, 53*(2), 123–151.
Wright, L. (1973). Functions. *Philosophical Review, 82*, 139–168.
Wright, L. (1976). *Teleological explanations: an etiological analysis of goals and functions.* Berkeley: University of California Press.
Young, K. D. (2006). The selective value of bacterial shape. *Microbiology and Molecular Biology Reviews, 70*(3), 660–703.
Zaretzky, A., & Letelier, J. C. (2002). Metabolic networks from (M,R) systems and autopoiesis perspective. *Journal of Biological Structures, 10*(3), 265–280.
Zepik, H., Blöchliger, E., & Luisi, P. L. (2001). A chemical model of homeostasis *Angew. Chemie Int, 40*(1), 199–202.
Ziemke, T. (2003). What's that thing called embodiment? In *Proceedings of the 25th Annual Meeting of the Cognitive Science Society.*
Zullo, L., Sumbre, G., Agnisola, C., Flash, T., & Hochner, B. (2009). Nonsomatotopic organization of the higher motor centers in octopus. *Current Biology, 19*, 1632–1636.

Index

A. Moreno, M. Mossio, *Biological Autonomy*, History, Philosophy and Theory of the Life Sciences 12, DOI 10.1007/978-94-017-9837-2

MIX
Papier aus verantwortungsvollen Quellen
Paper from responsible sources
FSC® C105338

If you have any concerns about our products,
you can contact us on
ProductSafety@springernature.com

In case Publisher is established outside the EU,
the EU authorized representative is:
Springer Nature Customer Service Center GmbH
Europaplatz 3, 69115 Heidelberg, Germany

Printed by Libri Plureos GmbH
in Hamburg, Germany